Biochemistry

Basic Sciences:
PreTest Self-Assessment and Review Series

John Watkins Foster, Jr., M.D., Editor
Yale-New Haven Medical Center
New Haven, Connecticut

Robert Edward Humphreys, M.D., Ph.D., Editor
University of Massachusetts Medical School
Worcester, Massachusetts

Biochemistry

PreTest Self-Assessment and Review

Edited by
John M. Kirkwood, M.D.
Yale-New Haven Medical Center
New Haven, Connecticut

PreTest Service, Inc.,
Distributed by Blakiston Publications
McGraw-Hill Book Company

PreTest Service, Inc., Wallingford 06492

Editor: Mary Ann C. Sheldon
Typographer: Elaine Reid
Production Staff: Donna D'Amico,
Pamela G. Oliano, Judith M. Raccio
Layout: Robert Tutsky
Illustrator: Leonard Galushko
Cover Design: Silverman Design
Printer: Connecticut Printers

Library of Congress Catalog Card Number: 76-2485
ISBN: 0-07-050796-1

NOTICE

Medicine is an ever-changing science. As new research and clinical experience broaden our knowledge, changes in treatment and drug therapy are required. The editors and the publisher of this work have made every effort to ensure that the drug dosage schedules herein are accurate and in accord with the standards accepted at the time of publication. Readers are advised, however, to check the product information sheet included in the package of each drug they plan to administer to be certain that changes have not been made in the recommended dose or in the contraindications for administration. This recommendation is of particular importance in regard to new or infrequently used drugs.

Contents

Part One: Questions

Part Two: Answers, Explanations, and References

List of Contributors

Jane Barry, M.D.
Yale-New Haven Medical Center
New Haven, Connecticut

Stewart Fox, M.D.
Yale-New Haven Medical Center
New Haven, Connecticut

Stephen Gray, M.D.
U. S. Army Hospital
Landstuhl, West Germany

Donald Francis Haggerty, Ph.D.
University of California, Los Angeles
Los Angeles, California

T. Ralph Hollands, M.D., Ph.D.
McMaster University
Hamilton, Ontario, Canada

John T. Homer, Ph.D.
University of Oklahoma
Stillwater, Oklahoma

Thomas Johnson, M.D.
Yale-New Haven Medical Center
New Haven, Connecticut

L. James Kennedy, Jr., M.D.
University of Colorado,
School of Medicine
Denver, Colorado

John Kirkwood, M.D.
Yale-New Haven Medical Center
New Haven, Connecticut

John Kozarich, Ph.D.
Harvard University
Cambridge, Massachusetts

Ian Love, M.D.
Yale-New Haven Medical Center
New Haven, Connecticut

Lore Ann McNicol, Ph.D.
University of Pennsylvania
School of Medicine
Philadelphia, Pennsylvania

Monique Minor, M.D.
Yale-New Haven Medical Center
New Haven, Connecticut

Ronald Daniel Neumann, M.D.
Yale-New Haven Medical Center
New Haven, Connecticut

David A. Pearson, Ph.D.
Yale-New Haven Medical Center
New Haven, Connecticut

Duane L. Peavy, Ph.D.
Harvard Medical School
Boston, Massachusetts

Mary Lake Polan, M.D., Ph.D.
Yale-New Haven Medical Center
New Haven, Connecticut

William Schoene, M.D.
Harvard Medical School
Boston, Massachusetts

Richard Staimen, M.D.
Fort Devens Army Hospital
Ayre, Massachusetts

Daniel Carl Sullivan, M.D.
Yale-New Haven Medical Center
New Haven, Connecticut

Gary Boyd Thurman, Ph.D.
University of Texas Medical Branch
Galveston, Texas

Prabhakar Narhari Vaidya, M.D.
Oak Forest Hospital
Oak Forest, Illinois

Richard Weiss, Ph.D.
University of California
Irvine, California

Stephen White, Ph.D.
University of Chicago
Chicago, Illinois

Biochemistry

PreTest Self-Assessment and Review

Abbreviations

A	adenine
Acetyl CoA	acetyl coenzyme A
ADP	adenosine diphosphate
AMP	adenosine 5'-monophosphate (adenylic acid)
ATP	adenosine triphosphate
ATPase	adenosine triphosphatase
C	cytosine
CDP	cytidine diphosphate
CMP	cytidine monophosphate (cytidylic acid)
CoA	coenzyme A
CTP	cytidine triphosphate
Cyclic AMP	adenosine 3', 5'-cyclic monophosphate (3', 5'-cyclic adenylic acid)
DNA	deoxyribonucleic acid
DNase	deoxyribonuclease
Dopa	3,4-dihydroxyphenylalanine
FAD (FADH$_2$)	flavin adenine dinucleotide (reduced form)
FSH	follicle-stimulating hormone
G	guanine
GDP	guanosine diphosphate
GMP	guanosine 5'-monophosphate (guanylic acid)
GTP	guanosine triphosphate
IMP	inosine 5'-monophosphate (inosinic acid)
LH	luteinizing hormone
mRNA	messenger RNA
NAD⁺(NADH)	nicotinamide adenine dinucleotide (reduced form)
NADP⁺(NADPH)	nicotinamide adenine dinucleotide phosphate (reduced form)
P$_i$	inorganic phosphate
RNA	ribonucleic acid
RNase	ribonuclease
rRNA	ribosomal RNA
T	thymine
TDP	thymidine diphosphate
tRNA	transfer ribonucleic acid
TSH	thyroid-stimulating hormone
U	uracil
UDP	uridine diphosphate
UTP	uridine triphosphate

Amino Acids, Proteins, and Enzymes

DIRECTIONS: Each question below contains five suggested answers. Choose the **one best** response to each question.

1. The structural component of proteins that makes the greatest contribution to their optical absorbance at 280 nm is the

(A) indole ring of tryptophan
(B) phenol ring of tyrosine
(C) benzyl ring of phenyl-alanine
(D) sulfur atom of cysteine
(E) peptide bond

2. Rotation of polarized light is caused by solutions of all the following amino acids EXCEPT

(A) alanine
(B) glycine
(C) leucine
(D) serine
(E) valine

3. All of the following statements about the peptide glu-his-arg-val-lys-asp are true EXCEPT

(A) it migrates toward the anode at pH 12
(B) it migrates toward the cathode at pH 3
(C) it migrates toward the cathode at pH 5
(D) it migrates toward the cathode at pH 11
(E) its isoelectric point is approximately pH 8

4. The graph below shows a titration curve of a common biochemical compound. All of the following statements about the graph are true EXCEPT

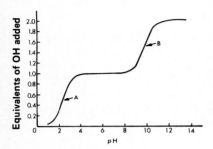

(A) the compound has two ionizable functions
(B) the compound is a simple amino acid
(C) the maximum buffering capacity of the compound is between pH 5 and 7
(D) point A could represent the range of ionization of a carboxyl function
(E) points A and B represent the respective pKs of an acidic and basic function

5. Which amino acid has an ionizable side chain pK closest to physiologic pH responsible for the major buffering effect of proteins?

(A) Cysteine
(B) Glutamic acid
(C) Glutamine
(D) Histidine
(E) Lysine

6. Which of the following amino acids has a net positive charge at physiologic pH?

(A) Cysteine
(B) Glutamic acid
(C) Lysine
(D) Tryptophan
(E) Valine

7. The concentration of hydrogen ions in a solution is measured by the pH, that numerically is equivalent to

(A) $\log_{10} (H^+)$
(B) $-\log_{10} (H^+)$
(C) $\log_e (H^+)$
(D) $-\log_e (H^+)$
(E) $1/\log_{10}(H^+)$

8. Ninhydrin reacts with amino acids by causing

(A) dehydrogenation
(B) a reduction of the amino group
(C) cleavage of certain peptide bonds
(D) cleavage of the side chain
(E) oxidative decarboxylation

9. Tyrosine may be detected in a mixture of amino acids by the

(A) Sakaguchi reaction (alpha-naphthol and sodium hypochlorite)
(B) nitroprusside reaction
(C) Ehrlich's reaction (p-dimethylaminobenzaldehyde)
(D) Millon reaction [$Hg(NO_3)_2$ in HNO_3]
(E) Sullivan reaction

10. Which of the following amino acids is synthesized only after incorporation of a precursor into a polypeptide?

(A) Proline
(B) Lysine
(C) Hydroxyproline
(D) Glutamate
(E) Serine

11. Which of the following amino acids is formed by transamination of a member of the citric acid cycle?

(A) Alanine
(B) Lysine
(C) Serine
(D) Tyrosine
(E) Valine

12. Which of the following amino acids can be metabolized to fatty acids in mammals?

(A) Isoleucine
(B) Methionine
(C) Arginine
(D) Histidine
(E) Tryptophan

13. Thyroxine is a derivative of

(A) threonine
(B) tryptophan
(C) tyrosine
(D) thiamine
(E) tyramine

14. Which of the following amino acids is NOT essential in mammals?

(A) Phenylalanine
(B) Lysine
(C) Tyrosine
(D) Leucine
(E) Methionine

15. Aspartic acid is incorporated in the synthesis of

(A) porphyrins
(B) steroids
(C) sphingolipids
(D) pyrimidines
(E) coenzyme A

Questions 16-17

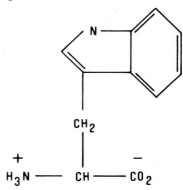

16. The figure shown above is

(A) coenzyme A
(B) NAD
(C) tyrosine
(D) histidine
(E) tryptophan

17. The compound shown above is found in

(A) DNA
(B) mRNA
(C) hemoglobin
(D) sphingomyelin
(E) the citric acid cycle

18. The only amino acid that is ketogenic but not glycogenic is

(A) isoleucine
(B) tyrosine
(C) leucine
(D) phenylalanine
(E) lysine

19. The excessive administration of which of the following amino acids would be expected to increase the excretion of ketones?

(A) Arginine
(B) Histidine
(C) Leucine
(D) Proline
(E) Valine

20. In the blood, the amino acid found in the highest concentration is

(A) alanine
(B) glutamine
(C) glycine
(D) histidine
(E) tryptophan

21. Phenylalanine and tyrosine enter the citric acid cycle after degradation to

(A) pyruvate
(B) fumarate
(C) succinyl-CoA
(D) α-ketoglutarate
(E) citrate

22. N^5-Methyltetrahydrofolate is a methylating agent that transfers a methyl group to

(A) acetate
(B) homocysteine
(C) norepinephrine
(D) pyruvic acid
(E) testosterone

23. S-Adenosylmethionine is a methylating agent that transfers a methyl group to

(A) acetate
(B) homocysteine
(C) norepinephrine
(D) pyruvic acid
(E) testosterone

24. Serine is converted to ethanolamine by the removal of

(A) hydrogen
(B) oxygen
(C) carbon dioxide
(D) ammonia
(E) a carboxyl group

25. Gamma decarboxylation of aspartic acid produces

(A) alanine
(B) asparagine
(C) glutamic acid
(D) glycine
(E) serine

26. A sulfur-containing amino acid that is NOT found in proteins is

(A) homocysteine
(B) homoserine
(C) methionine
(D) proline
(E) threonine

27. The decarboxylation of which of the following amino acids produces a vasodilating compound?

(A) Arginine
(B) Aspartic acid
(C) Histidine
(D) Glutamine
(E) Proline

28. In urea synthesis, which of the following enzymes uses adenosine triphosphate (ATP)?

(A) Arginase
(B) Transaminase
(C) Condensing enzyme
(D) Argininosuccinase
(E) Fumarase

29. In the cycle shown below, compound B is

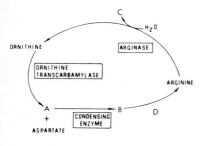

(A) pyruvate
(B) urea
(C) malate
(D) lactate
(E) argininosuccinate

30. Lysine would be most likely to inhibit which of the following reactions in *E. coli*?

(A) Aspartic acid ⇌ β-aspartyl phosphate
(B) α-Ketoisovaleric acid ⇌ valine
(C) Threonine ⇌ acetaldehyde + glycine
(D) α-Ketoadipic acid ⇌ γ-aminolevulinic acid
(E) None of the above reactions

31. All the following statements about the α-helix are true EXCEPT

(A) it is stabilized by intramolecular hydrogen bonds
(B) it is stabilized by minimizing unfavorable R-group interactions
(C) it is stabilized by hydrophobic interactions
(D) prolyl and glycyl residues tend to interrupt α-helical structure
(E) it is one type of secondary structure found in some proteins

32. The differences seen among the isozymes of a given enzyme are caused by

(A) variations in the assembly of a number of different monomeric subunits into distinct multimeric complexes
(B) variations in substrate specificity
(C) variations in the catalytic efficiency
(D) immunologic variations based on differences in protein structure
(E) the ability of a given enzyme to function differently in a variety of physiologic environments

A   B   C   D   E

33. Which of the schematic drawings of protein configurations shown above represents a super coiled helix?

(A) Figure A
(B) Figure B
(C) Figure C
(D) Figure D
(E) Figure E

34. Figure D is a three-stranded structure which could represent the conformation of

(A) hemoglobin
(B) α-keratin
(C) polylysine
(D) silk fibroin
(E) tropocollagen

---

35. Lactate dehydrogenase (LDH) is a tetramer composed of two different polypetide chains. Assuming these chains associate at random to form the enzyme, how many isozymes does this enzyme possess?

(A) 2
(B) 3
(C) 4
(D) 5
(E) 6

36. All of the following gastrointestinal enzymes are secreted as inactive zymogens (proenzymes) EXCEPT

(A) pepsin
(B) carboxypeptidase
(C) chymotrypsin
(D) trypsin
(E) ribonuclease

37. Which enzyme has the greatest specificity for peptide·bonds on the carboxyl side of a cationic amino acid side chain?

(A) Carboxypeptidase
(B) Chymotrypsin
(C) Pepsin
(D) Rennin
(E) Trypsin

38. Myosin is

(A) a zinc-requiring enzyme
(B) a spherically symmetric molecule
(C) a cyclic adenosine 5'-mono-phosphate (AMP) phosphodiesterase
(D) an actin-binding protein
(E) low in α-helix content

39. Immunoglobulins

(A) contain no carbohydrate
(B) can be separated into different classes by electrophoresis
(C) are always at a constant serum level
(D) are linked between the heavy and light chains by nucleic acids
(E) are synthesized by the mucosal cells of the appendix

40. The first step in the catabolism of hemoglobin occurs when hemoglobin is

(A) converted to biliverdin in the liver
(B) opened in the reticuloendothelial cells
(C) converted to bilirubin in the reticuloendothelial cells
(D) conjugated with glucuronic acid in the liver
(E) reduced in the liver

41. The normal brown-red color of feces is caused by their content of

(A) stercobilin
(B) urobilinogen
(C) bilirubin
(D) mesobilirubin
(E) biliverdin

42. K_m and V_{max} can be determined from the Lineweaver-Burk plot of the Michaelis-Menten equation shown below. When V is the reaction velocity at substrate concentration S, the abscissa experimental data should be

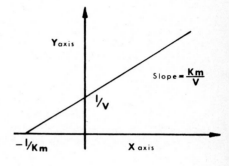

(A) 1/V
(B) V
(C) 1/S
(D) S
(E) V/S

43. If an enzyme is only active when a particular histidyl group is not protonated, then increasing the proton concentration, i.e., decreasing pH, would give which of the following types of inhibition pattern?

(A) Uncompetitive
(B) Noncompetitive
(C) Competitive
(D) Mixed
(E) None of the above

44. A purely competitive inhibitor of an enzyme

(A) increases K_m without affecting V_{max}
(B) decreases K_m without affecting V_{max}
(C) increases V_{max} without affecting K_m
(D) decreases V_{max} without affecting K_m
(E) decreases both V_{max} and K_m

45. Enzymes as classic catalysts

(A) raise the energy of activation
(B) lower the energy of activation
(C) raise the energy level of the products
(D) lower the energy level of the reactants
(E) decrease the free energy of the reaction

46. The K_m of the enzyme giving the kinetic data shown below is

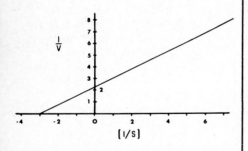

(A) 0.11
(B) 0.25
(C) 0.33
(D) -0.25
(E) -0.50

47. Given that $\Delta F^0 = -RT(2.3) \log K_{eq}$, determine the free energy of the following reaction.

A	+	B	⇌	C
10		10		10
moles		moles		moles

(A) 2.3 RT
(B) −2.3 RT
(C) 4.6 RT
(D) −4.6 RT
(E) −9.2 RT

48. If curve A shown below represents no inhibition for the reaction of some enzyme with its substrate, which of the curves below would represent competitive inhibition of the same reaction?

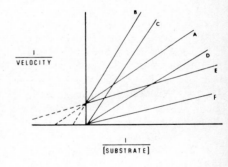

(A) B
(B) C
(C) D
(D) E
(E) F

8

49. If an enzyme behaves according to classic Michaelis-Menten kinetics, from a plot of the reciprocal velocity versus the reciprocal substrate concentration, the value for the Michaelis constant of the substrate can be determined graphically as the

(A) point of inflection of the curve
(B) slope of the line
(C) value of the intercept of the line with the ordinate
(D) absolute value of the intercept of the line with the abscissa
(E) reciprocal of the absolute value of the intercept of the line with the abscissa

50. The overall energy changes in biochemical reactions are

(A) influenced by the energy barrier of the reaction
(B) altered by cofactors
(C) proportionate to the concentration of the reactants
(D) essentially zero at equilibrium
(E) independent of the mechanism of the reaction

51. Which of the following oxidation-reduction systems has the highest redox potential?

(A) Fumarate/succinate
(B) Ubiquinone ox./red.
(C) Fe^{+++} cytochrome a/Fe^{++}
(D) Fe^{+++} cytochrome b/Fe^{++}
(E) NAD^+/NADH

52. The second law of thermodynamics states that

(A) perpetual motion is theoretically attainable at $O°K$
(B) energy and mass are conserved and interchangeable
(C) in an energetically closed system any process goes spontaneously from a state of lesser order to a state of greater order
(D) in an energetically closed system any process goes spontaneously in the direction of increased entropy
(E) any system changes spontaneously in the direction of decreased free energy

53. An uncoupling agent of oxidative phosphorylation causes

(A) the ATPase activity of the mitochondria to decrease
(B) cellular respiration to cease
(C) mitochondrial oxidation of citric acid to cease
(D) ATP formation to stop and respiration to continue
(E) an increased utilization of intracellular glucose

54. Dinitrophenol would be most likely to inhibit cell function by disrupting

(A) glycolysis
(B) glyconeogenesis
(C) oxidative phosphorylation
(D) the citric acid cycle
(E) none of the above

9

55. The activity of most single polypeptide enzymes can be represented by the hyperbolic curve A shown below. However, the activity of homotropic regulatory enzymes shows a sigmoid dependence on substrate concentration, curve B. This sigmoid relationship between substrate concentration and reaction velocity indicates that

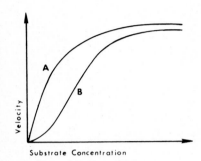

Substrate Concentration

(A) homotropic enzymes are polymers
(B) homotropic enzymes must catalyze several separate reactions on the way to the final product
(C) homotropic enzymes catalyze reactions more slowly than single polypeptide enzymes
(D) the reaction rate is independent of substrate concentration
(E) the binding of one substrate molecule enhances subsequent substrate binding and activity

56. Polymyxin is unique among chemotherapeutic agents because it is bactericidal in the absence of cell growth. It exerts its effect by

(A) binding to DNA as an insertion mutation
(B) binding to DNA polymerase
(C) binding to polysome-bound mRNA
(D) detergent-like disruption of membranes
(E) dissociating sigma from RNA polymerase

57. The action of cycloserine in inhibiting transpeptidation in the formation of the petidoglycan cell wall network of gram-positive organisms is inhibited competitively by

(A) D-alanine
(B) diaminopimelic acid
(C) D-glutamine
(D) L-lysine
(E) D-serine

58. Microscopically, tetracyclines localize within the

(A) nucleus
(B) mitochondria
(C) endoplasmic reticulum
(D) cell membrane
(E) Golgi apparatus

59. Which of the following types of enzymes does NOT have a demonstrable inducer?

(A) Allosteric enzyme
(B) Constitutive enzyme
(C) Isozymic enzyme
(D) Inhibited enzyme
(E) Cooperative enzyme

60. De novo synthesis of an enzyme, promoted by the substrate on which it acts, occurs in the process called

(A) activation
(B) derepression
(C) gratuity
(D) induction
(E) constitutivity

61. Control of metabolic paths may be exerted by enzyme repression or induction. In vertebrates this form of enzyme control occurs primarily in the

(A) heart
(B) skeletal muscle
(C) brain
(D) liver
(E) bone

62. A hypothetical biosynthetic pathway is shown in the diagram below. A microbial organism defective in one enzyme of this path is grown in a medium containing X. Large amounts of M and L are found in the organism but none of Z. In which enzyme is the mutation expressed?

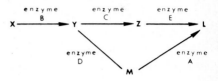

(A) Enzyme A
(B) Enzyme B
(C) Enzyme C
(D) Enzyme D
(E) Enzyme E

11

DIRECTIONS: Each question below contains four suggested answers of which **one** or **more** is correct. Choose the answer

A	if	**1, 2, and 3**	are correct
B	if	**1 and 3**	are correct
C	if	**2 and 4**	are correct
D	if	**4**	is correct
E	if	**1, 2, 3, and 4**	are correct

63. Which of the following statements about most peptide bonds between L-amino acids in proteins are true?

(1) The peptide bond has partial double bond character
(2) The peptide bond is shorter than a normal carbon-carbon single bond
(3) The peptide bond usually has a *trans* configuration with respect to the α-carbons of the two amino acids involved in the peptide bond
(4) There is perfectly free rotation about the peptide bond

64. Which of the following amino acids have relatively nonpolar, hydrophobic side chains?

(1) Isoleucine
(2) Methionine
(3) Proline
(4) Phenylalanine

65. Which of the following statements about amino acid biosynthesis are true?

(1) Essential amino acids are required in an animal's diet to maintain a proper nitrogen balance
(2) Methionine is an essential amino acid for humans
(3) The "essentiality" of an amino acid can vary as a function of physiologic state, e.g., pregnancy
(4) Histidine is an essential amino acid for humans

66. In mammalian tissues, glycine is a biosynthetic precursor of

(1) heme
(2) creatine
(3) guanine
(4) thymine

67. In mammalian tissues, serine can be a biosynthetic precursor of

(1) methionine
(2) glycine
(3) tryptophan
(4) choline

68. Penicillin may be considered to be a condensation product of

(1) alanine
(2) cysteine
(3) isoleucine
(4) valine

69. D-alanine, illustrated below, is a structural analog of

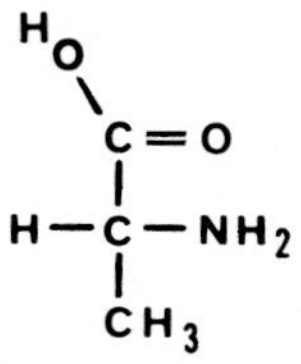

$$H-\overset{\overset{\displaystyle H}{\underset{\displaystyle O}{|}}}{\underset{\overset{\displaystyle |}{C}H_3}{\underset{|}{C}}}-NH_2$$

(1) streptomycin
(2) vancomycin
(3) bacitracin
(4) cycloserine

70. Which of the following drugs work by decreasing the levels of folate cofactors in bacteria?

(1) p-Aminosalicylic acid (PAS)
(2) Isoniazid (INH)
(3) Sulfisoxazole (Gantrisin)
(4) Tetracycline

71. Which of the following statements about the structure of proteins are NOT correct?

(1) Intrachain disulfide bonds may not be crucial in predetermining the conformation of a protein molecule
(2) Charged amino acid side chains tend to be on the outside of the molecule exposed to solvent
(3) The primary structure of proteins is among the more important factors in determining the higher order structure
(4) Hydrophobic side chains of amino acid residues are only rarely buried in the center of molecules hidden from aqueous solvent

72. Which of the following statements about gamma immunoglobulins are correct?

(1) There are two antigen binding sites per antibody molecule
(2) In multiple myeloma, incomplete immunoglobulins may be excreted in the urine (Bence-Jones proteins)
(3) Both heavy and light immunoglobulin chains have constant C-terminal and variable N-terminal sequences
(4) The only chemical forces holding immunoglobulin protein chains together are non-covalent in nature

73. Which of the following statements concerning trypsinogen and chymotrypsinogen are false?

(1) They have considerable homologies in primary sequence
(2) They can be converted into active enzymes by limited digestion with trypsin
(3) They are secreted by exocrine cells in the pancreas
(4) They are exopeptidases

74. Most enzymes

(1) increase the rapidity of the reaction they catalyze
(2) are specific for the substrate as well as the reaction catalyzed
(3) are large polypeptides with a mass greater than 5000 daltons
(4) are most active near neutral pH

75. The figure below shows the structure of an immunoglobulin hydrolyzed by papain to form two A fragments and one B fragment. Fragment A

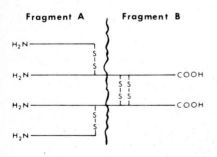

(1) contains the antibody combining site
(2) is Bence-Jones protein
(3) contains the "hypervariable region" sequences
(4) is the heavy chain

76. Proteins that contain a porphyrin ring include

(1) hemoglobin
(2) myoglobin
(3) cytochrome
(4) catalase

77. Which of the following proteins contain iron?

(1) Cytochrome c
(2) Hemoglobin
(3) Myoglobin
(4) Peroxidase

78. The complement system

(1) is composed of a group of soluble globulins
(2) produces peptides with chemotactic activity
(3) is lytic to red cells
(4) usually requires only a bacterium to initiate its action

79. The Monod model for allosteric enzymes assumes

(1) all allosteric enzymes are polymers
(2) each subunit bears both a catalytic and an allosteric site
(3) the binding affinity of a subunit for a ligand may differ according to the conformational state of the subunit
(4) transitions from state to state involve changes in all subunits

Questions 80-81

The velocity-substrate curve below characterizes an allosteric enzyme system.

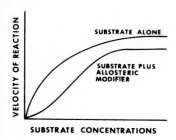

80. Which of the following statements are true?

(1) A modifier present at the allosteric site can also affect the catalytic site
(2) A modifier changes the binding constant for the substrate, without changing the velocity of the reaction
(3) The binding of substrate is dependent on its concentration
(4) The binding of modifier is independent of its concentration

81. This type of allosteric behavior is demonstrated by

(1) phosphoglucose isomerase
(2) aspartate transcarbamoylase
(3) lactate dehydrogenase
(4) hemoglobin

82. A good measure of the degree of order in a biologic system or molecular constituent is

(1) the equilibrium constant for the formation of the system or molecule, K_{eq}
(2) the enthalpy of formation for the system or molecule, ΔH°
(3) the free energy of formation for the system or molecule, ΔF°
(4) the entropy change associated with the formation of the system or molecule, ΔS°

83. Which of the following statements about hemoglobin transport of O_2 are FALSE?

(1) Each of the four heme moieties binds O_2 independently
(2) The graph of percent O_2 bound versus O_2 pressure is sigmoid in shape
(3) The association constant for the interaction of O_2 with hemoglobin is greater than that for the interaction of CO with hemoglobin
(4) The binding of O_2 to hemoglobin causes no valence change in the iron of the heme moiety

84. Which of the following hypotheses are currently being employed to explain oxidative phosphorylation?

(1) Chemical coupling hypothesis
(2) Warburg hypothesis
(3) Chemiosmotic hypothesis
(4) Pasteur effect

15

SUMMARY OF DIRECTIONS

A	B	C	D	E
1, 2, 3 only	1, 3 only	2, 4 only	4 only	All are correct

85. Which of the following statements are true of oxidation-reduction potentials?

(1) The common standard, potential = 0, is arbitrarily assigned to the H_2 electrode
(2) pH has no relation to oxidation-reduction potentials
(3) Free energy changes may be calculated from oxidation-reduction potentials
(4) Metallic electrodes are required for determination of oxidation-reduction potentials

86. Properties of the b cytochromes of the mitochondrial electron transport chain include

(1) a standard redox potential lower than that of cytochromes c and a
(2) ready detachment from mitochondrial membranes
(3) low reactivity with cyanide or CO, which are inhibitors of some hemoproteins
(4) ease of reaction with cytochrome a

87. Phosphorylation sites in the mitochondrial respiratory chain may occur between

(1) coenzyme Q and cytochrome b
(2) cytochrome b and cytochrome c
(3) pyruvate and NAD
(4) NAD and a flavoprotein

88. Sickle cell anemia

(1) results from a single amino acid change in hemoglobin
(2) is caused by a phospholipid change in erythrocytes
(3) is seen in homozygous individuals only
(4) is widely distributed in the sonoran life zone

89. Synthesis of which of the following proteins is inducible?

(1) β-Galactoside permease of *B. subtilis*
(2) β-Lactamase in *S. aureus*
(3) Glucose-phosphate transferase of *E. coli*
(4) Immunoglobulins in cat lymphocytes

90. Allosteric inhibitor enzymes

(1) may produce effects proportional to the square of their concentration
(2) have only one catalytic site
(3) can have their effects reversed by substrate analogs
(4) are not affected by high salt denaturation

91. Factors circulating in the bloodstream that are involved in the formation of blood clots include

(1) plasmin
(2) fibrinogen
(3) heparin
(4) platelets

92. The active transport of β-galactosides in *E. coli* is characterized by

(1) a requirement for an energy source
(2) a saturating concentration of galactoside above which a more rapid rate of uptake cannot be obtained
(3) a rate of efflux dependent on the intracellular concentration of galactoside
(4) being inducible in cells grown solely in lactose as an energy source

93. Heavy chains of IgG antibody may be separated from light chains with

(1) ethanolamine
(2) pepsin
(3) papain
(4) mercaptoethanol

DIRECTIONS: The groups of questions in this section consist of five lettered headings followed by several numbered items. For each numbered item choose the **one** lettered heading with which it is **most** closely associated. Each lettered heading may be used once, more than once, or not at all.

Questions 94-95
For each amino acid, choose the compound from which it is synthesized.

(A) Glutamate
(B) Aspartate
(C) Ribose 5-phosphate
(D) Shikimic acid
(E) Serine

94. Histidine

95. Proline

Questions 96-99
For each amino acid, choose the appropriate description of its side chain.

(A) Acidic
(B) Basic
(C) Aromatic
(D) Sulfur-containing
(E) Branched chain aliphatic

96. Cysteine

97. Lysine

98. Leucine

99. Phenylalanine

Questions 100-102
For each amino acid, choose the compound in which it appears most characteristically.

    (A)   Chondroitin sulfate
    (B)   Collagen
    (C)   Keratin
    (D)   Melanin
    (E)   Myosin

100. Cystine

101. Tyrosine

102. Hydroxyproline

Questions 103-104
For each chemical modification seen in proteins, choose the reagent that causes that effect.

    (A)   Chymotrypsin
    (B)   Trypsin
    (C)   2,4-Dinitrophenol (DNP)
    (D)   Mercaptoethanol
    (E)   Phenylisothiocyanate

103. Cleavage of protein disulfide bridges

104. Hydrolysis of peptide linkages containing aromatic amino acid residues

Questions 105-107
For each pancreatic enzyme, choose the description of its principal specificity as a peptidase.

    (A)   Cleaves after proline
    (B)   Cleaves after methionine
    (C)   Cleaves after aromatic amino acids
    (D)   Cleaves after lysine and arginine
    (E)   Exopeptidase

105. Trypsin

106. Chymotrypsin

107. Carboxypeptidase

Questions 108-110
In the following enzyme-catalyzed reaction E, S, and P are the enzyme, substrate, and product, respectively. (E_0 is the total enzyme concentration.) k_1, k_2, and k_3 are rate constants.

$$S + E \underset{k_2}{\overset{k_1}{\rightleftharpoons}} S{\cdot}E \overset{k_3}{\Rightarrow} P + E$$

For each kinetic parameter, choose the equivalent expression in terms of the rate constants.

    (A) $k_3 E_0$
    (B) k_2/k_3
    (C) k_2/k_1
    (D) $(k_2 + k_3)/k_1$
    (E)   None of the above

108. V_{max}

109. K_s (dissociation constant for E·S complex)

110. K_m (Michaelis constant)

Nucleic Acids and Molecular Genetics

DIRECTIONS: Each question below contains five suggested answers. Choose the **one best** response to each question.

111. Only DNA, and not RNA, would be radioactively labeled in an animal fed tritiated

(A) adenine
(B) cytosine
(C) guanine
(D) thymine
(E) uracil

112. Which of the following nucleic acid bases is found in mRNA but not in DNA?

(A) Adenine
(B) Cytosine
(C) Guanine
(D) Uracil
(E) Thymine

113. Unusual nucleotide bases are found primarily in

(A) rRNA
(B) mRNA
(C) tRNA
(D) nucleolar DNA
(E) mitochondrial DNA

114. The absence of which enzyme involved in the "salvage pathways" of nucleotide metabolism results in a severe hereditary form of gout?

(A) Hypoxanthine-guanine phosphoribosyltransferase
(B) Aspartate transcarbamoylase
(C) Thymidylate kinase
(D) Adenylate deaminase
(E) Xanthine oxidase

115. Which of the following compounds does NOT receive a methyl group from S-adenosylmethionine?

(A) Creatine phosphate
(B) Epinephrine
(C) Melatonin
(D) Phosphatidylcholine
(E) Thymine

116. S-Adenosylmethionine is shown below with five substituent groups labeled A through E. Which group is S-adenosylmethionine able to donate in purine biosynthesis?

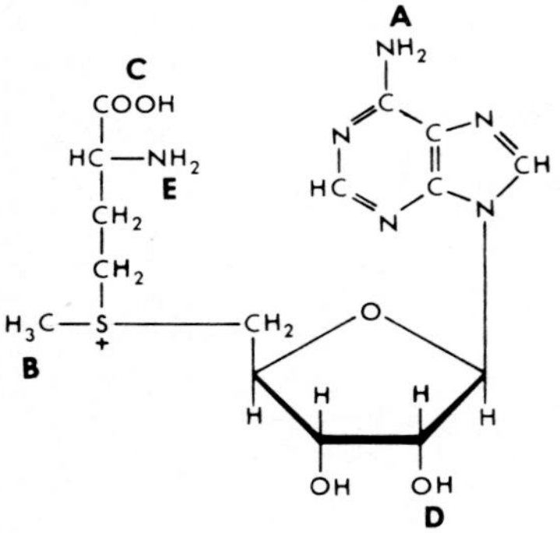

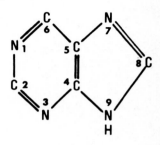

(A) A
(B) B
(C) C
(D) D
(E) E

117. The end product of purine metabolism that is excreted by man is

(A) allantoic acid
(B) orotic acid
(C) urea
(D) uric acid
(E) xanthine

118. Which of the following is a metabolic pathway common to bacteria and man?

(A) Purine synthesis
(B) Nitrogen fixation
(C) Cell wall mucopeptide synthesis
(D) Noncyclic photophosphorylation
(E) Fermentation to ethyl alcohol

119. Carbons 4 and 5 in the purine nucleus shown below are obtained from

(A) acetate
(B) alanine
(C) glycine
(D) aspartate
(E) glutamate

20

120. Which of the following statements about the lettered ring constituents of uracil shown below is true?

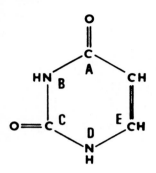

(A) The carbon atom lettered A is derived from the carboxyl side chain of aspartic acid
(B) The nitrogen atom lettered B is derived from the ε-amino group of lysine
(C) The carbon atom lettered C is derived from the carboxyl of methionine
(D) The nitrogen atom lettered D is derived from the amino group of glutamine
(E) The carbon atom lettered E is donated by folic acid as a free methyl group

121. Which of the following is true of de novo pyrimidine synthesis but not of purine biosynthesis?

(A) It is synthesized attached to ribose 5-phosphate
(B) Carbon-1 fragments are donated by folic acid derivatives
(C) Carbamoyl phosphate donates a carbamoyl group
(D) Glycine is incorporated intact
(E) Glutamine is a nitrogen donor

122. Which of the following statements is true of the mitochondrial carbamoyl phosphate synthetase but not of the cytoplasmic enzyme?

(A) It is inhibited by uridine triphosphate (UTP)
(B) It is involved in pyrimidine biosynthesis
(C) It is present in relatively low acitivity
(D) It is activated by acetylglutamate
(E) None of the above

123. Pyrimidine feedback inhibition controls the activity of

(A) dihydro-orotase
(B) orotidylic pyrophosphorylase
(C) reductase
(D) aspartate transcarbamoylase
(E) hydroxymethyl cytidylate synthetase

124. The degradation of RNA by pancreatic ribonuclease produces

(A) nucleoside 2'-phosphates
(B) nucleoside 3'-phosphates
(C) nucleoside 5'-phosphates
(D) nucleotides
(E) oligonucleosides

125. The sedimentation coefficient of a DNA molecule depends on all of the following EXCEPT

(A) partial specific volume
(B) diffusion coefficient
(C) optical density
(D) molecular weight
(E) molecular shape

126. The point which is called T_m (melting temperature) for double-stranded DNA is represented by what letter in the diagram below?

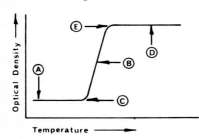

(A) A
(B) B
(C) C
(D) D
(E) E

127. Which of the following factors leads to an increased melting temperature for duplex DNA?

(A) High content of adenine (A) + guanine (G)
(B) High content of cytosine (C) + thymine (T)
(C) High content of adenine + thymine
(D) High content of cytosine + guanine
(E) None of the above

128. The Watson-Crick model of DNA structure shows

(A) a triple-stranded structure
(B) the DNA strands running in opposite directions
(C) pair bonding between bases A and G
(D) covalent bonding between bases
(E) the phosphate backbone to be on the inside of the DNA helix

129. Which of the following statements concerning histones is NOT true?

(A) They are of low molecular weight
(B) There are many (>1000) varieties of distinct histone molecules in the nucleus of a given cell type
(C) There are known interspecies homologies in the amino acid sequences of histones
(D) Histones are rich in arginine and lysine
(E) Histones are non-covalently attached to DNA in stoichiometric amounts

130. During DNA replication, the sequence 5'-TpApGpAp-3' would produce which of the following complementary structures?

(A) 5'-TpCpTpAp-3'
(B) 5'-ApTpCpTp-3'
(C) 5'-UpCpUpAp-3'
(D) 5'-GpCpGpAp-3'
(E) 3'-TpCpTpAp-3'

131. AUG, the only identified codon for methionine, is important as

(A) a chain-terminating codon
(B) the site of attachment for the 30 S ribosomal particle
(C) a chain-initiating codon
(D) a releasing factor for peptide chains
(E) the recognition site on the transfer RNA

132. The fact that DNA bears the genetic information of an organism implies that

(A) base composition should be identical from species to species
(B) viral infection is accomplished by transfer of protein into the host cell
(C) DNA from different tissues in the same organism should have the same base composition
(D) DNA base composition should change with age and nutritional state of the organism
(E) DNA occurs in small ring-shaped structures

133. DNA replicates in a semiconservative manner. If a completely radioactive double-stranded DNA molecule undergoes two rounds of replication in a solution free of radioactive label, of the resulting four DNA molecules

(A) half should contain no radioactivity
(B) all should contain radioactivity
(C) half should contain radioactivity in both strands
(D) one should contain radioactivity in both strands
(E) none should contain radioactivity

134. Which of the following mutations is most likely to be lethal?

(A) Substitution of adenine for cytosine
(B) Substitution of cytosine for guanine
(C) Substitution of methylcytosine for cytosine
(D) Deletion of three nucleotides
(E) Insertion of one nucleotide

135. Thymine dimers which occur in DNA as a mutational event

(A) do not stop replication
(B) are repaired by an enzyme system that includes the enzyme ligase
(C) are read as frame-shift mutations
(D) cause covalent bonds between thymines on opposite nucleotide strands
(E) are caused by the enzyme thymine dimerase

136. Sickle cell anemia is the clinical manifestation of homozygous genes for an abnormal hemoglobin molecule. The mutational event responsible for the mutation in the beta chain is

(A) crossing over
(B) insertion
(C) deletion
(D) nondisjunction
(E) point mutation

137. What is the anticodon in tRNA that corresponds to the codon ACG in mRNA?

(A) UGC
(B) TGC
(C) GCA
(D) CGU
(E) CGT

138. Which of the following amino acid residues is thought to be the amino terminus of all polypeptide chains at the time of synthesis in *E. coli*?

(A) Methionine
(B) Serine
(C) N-Formylmethionine
(D) N-Formylserine
(E) Glutamate

139. What is wobble?

(A) The ability of certain anti-codons to pair with codons that differ at the third base
(B) A mechanism that allows for peptide extension in the 50 S subunit of the ribosome
(C) An error in translation induced by streptomycin
(D) The induction of lysogenic phages to become virulent
(E) Thermal motions leading to local denaturation of the DNA double helix

140. How many codons are recognized to terminate polypeptide chain elongation in protein synthesis?

(A) 5
(B) 4
(C) 3
(D) 2
(E) 1

141. All of the following statements about DNA-primed RNA synthesis are true EXCEPT

(A) RNA polymerase catalyzes the formation of phosphodiester bonds only in the presence of DNA
(B) only one strand of DNA serves as template in most circumstances
(C) the direction of growth of the RNA chain is from 5'- to 3'-end
(D) the newly synthesized RNA remains attached to both double-stranded and single-stranded DNA templates in vitro
(E) the RNA chain synthesized is never circular

142. What is sigma factor?

(A) A subunit of RNA polymerase responsible for the specificity of the initiation of transcription of RNA from DNA
(B) A subunit of the 30 S ribosome to which mRNA binds
(C) A subunit of the 50 S ribosome which catalyzes peptide bond synthesis
(D) The factor that forms the bridge between the 30 S and 50 S particles which make up the 70 S ribosome
(E) A subunit of DNA polymerase which allows for bidirectional synthesis in both 5' to 3' and 3' to 5' directions

24

143. The first step in the utilization of amino acids for protein synthesis requires the

(A) arginyl-RNA synthetase
(B) aminoacyl soluble RNA synthetase
(C) amino acid activating enzyme
(D) chain initiating enzyme
(E) binding of N-formylmethionyl-tRNA to the ribosomes

144. Which of the following reactions requires GTP?

(A) The synthesis of peptides at ribosomal sites
(B) The activation of amino acids by aminoacyl-RNA synthetase
(C) The release of peptides from polyribosomes
(D) The activation of arginine by arginyl-RNA synthetase
(E) The formation of polyribosomal units

145. Puromycin has been useful in studying protein synthesis because it

(A) alters the binding of tRNA to mRNA
(B) binds to the 30 S subunit of the ribosome
(C) releases peptide material from ribosomal complexes
(D) inhibits the amino acid activating enzyme
(E) prevents the formation of polysomes

146. Tetracycline prevents synthesis of polypeptides by

(A) competing with mRNA for ribosomal binding sites
(B) blocking mRNA formation from DNA
(C) releasing peptides from mRNA-tRNA complexes
(D) inhibiting tRNA binding to mRNA
(E) preventing amino acid binding by tRNA

147. Chloramphenicol

(A) inhibits protein synthesis at the 50 S unit ribosomal site
(B) inhibits protein synthesis at the 30 S unit ribosomal site
(C) does not block antibody synthesis
(D) causes misreading of the mRNA
(E) depolymerizes DNA

148. 5-Bromouracil is mutagenic because

(A) it inserts extra base pairs during replication
(B) it produces "thymine dimers"
(C) when it is incorporated into DNA in place of thymine, codons are misread
(D) when it is incorporated into RNA in place of adenine, codons are misread
(E) it deaminates adenine, guanine, or cytosine

149. What is the product of translation?

(A) Protein
(B) tRNA
(C) mRNA
(D) DNA
(E) rRNA

150. Ribosomes are

(A) an integral part of transcription
(B) bound together so tightly they cannot dissociate under physiologic conditions
(C) found both free in the cytoplasm and bound to membranes
(D) composed of RNA, DNA, and protein
(E) composed of three unequally sized subunits

151. Which of the following antimetabolites inhibits the production of poly A-containing RNA?

(A) Pactomycin
(B) Actinomycin
(C) Puromycin
(D) Cordycepin
(E) Cycloheximide

152. Peptide chain elongation involves all of the following EXCEPT

(A) peptidyl transferase
(B) GTP
(C) Tu, Ts, and G factors
(D) formyl-Met-tRNA
(E) mRNA

153. The cell wall of gram-positive bacteria is

(A) not weakened by lysozyme
(B) less than 15 Å thick
(C) composed partly of N-acetyl-muramic acid
(D) characterized by a lipoprotein layer
(E) composed of only L-amino acids

154. In the process known as transformation, bacteria

(A) pass their DNA from cell to cell
(B) are transformed by a bacterial phage
(C) take up pieces of DNA from the media
(D) develop point mutations
(E) develop frame-shift mutations

155. The classic phage shown to cause specialized transduction in *E. coli* is

(A) φX 10
(B) T 4
(C) phage λ
(D) K 12
(E) P 22

156. The transfer of a fragment of donor chromosome to a recipient cell by a bacteriophage which has been produced in the donor cell is called

(A) transduction
(B) transformation
(C) conjugation
(D) integration
(E) mutation

157. The process of transferring a large piece of chromosomal material between donor and recipient microbial cells is known as

(A) generalized transduction
(B) restricted transduction
(C) transformation
(D) conjugation
(E) repression transfer

158. Cycloheximide does NOT inhibit protein synthesis in

(A) HeLa cells
(B) molds
(C) *Cryptococcus neoformans*
(D) *Saccharomyces cerevisciae*
(E) *Plasmodium falciparum*

159. Interferon

(A) is virus-specific
(B) inhibits viral multiplication in all cells
(C) is a synthetic antiviral agent
(D) is a bacterial product
(E) requires expression of cellular genes

160. The most likely cause of a spontaneous point mutation in an *E. coli* culture is

(A) a tautomeric shift of electrons
(B) a break in the sugar-phosphate backbone of DNA
(C) the insertion of a single base pair
(D) exposure to ultraviolet light
(E) exposure to nitrous acid

161. A frame-shift mutation caused by the insertion or deletion of a base pair is most likely to be caused by

(A) acridine derivatives
(B) 5-bromouracil
(C) azaserine
(D) ethyl ethanesulfonate (EES)
(E) azathioprine

162. An induced chromosomal mutation that consists of the deletion or insertion of a single DNA base pair, is most likely to be caused by

(A) nitrous acid
(B) proflavine
(C) ethylmethanesulfonate
(D) x-rays
(E) 5-bromouracil

163. The location on the lactose operon shown below where RNA polymerase and sigma factor bind initially before initiating transcription is

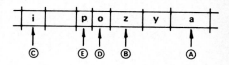

(A) A
(B) B
(C) C
(D) D
(E) E

164. The smallest unit of DNA capable of regulating the synthesis of a polypeptide is the

(A) operon
(B) cistron
(C) recon
(D) muton
(E) repressor gene

165. The electron micrograph below shows a circular, double-stranded molecule of DNA weighing ten million daltons. It is typical of DNA found in

(A) the nuclei of eukaryotes
(B) mammalian chromosomes
(C) *E. coli*
(D) mitochondria
(E) lysosomes

166. In a Hfr (high frequency of recombination) x F⁻ cross, the initiation point of chromosome transfer is determined by

(A) the genotype of the recipient F⁻ strain
(B) the genotype of the Hfr strain
(C) the integration of the F factor
(D) the conditions of the conjugation
(E) the phenotype of the Hfr strain

Photograph accompanies Question 165

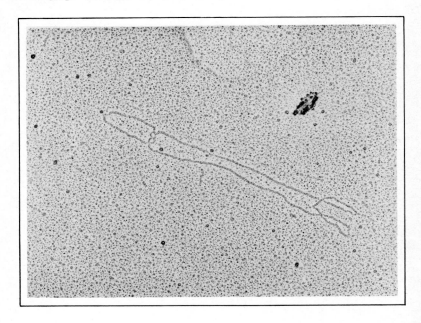

167. When the distance between genetic loci is sufficiently long to permit double cross-overs, the recombination frequency between the two genetic loci

(A) is always less than the sum of intervening frequencies
(B) is always more than the sum of intervening frequencies
(C) is equal to the sum of intervening frequencies
(D) exhibits a direct linear relationship with the cross-over frequency
(E) cannot be qualitatively related to the sum of intervening frequencies

168. The percentage of a bacterial chromosome usually involved in transformations is

(A) 1 percent
(B) 10 percent
(C) 25 percent
(D) 50 percent
(E) 100 percent

169. A common characteristic of conjugating bacteria is

(A) the transfer of the bacterial chromosome
(B) the formation of recombinants
(C) the conversion of the female cells to male cells
(D) the presence of F pili
(E) the presence of a high frequency recombinant

170. In the cross of *E. coli* strain $str^S F' lac^+$ with strain $lac^- str^r$ the desired recipients may be selectively grown on

(A) lactose minimal plates
(B) lactose minimal streptomycin plates
(C) glucose minimal streptomycin plates
(D) glucose minimal plates
(E) arabinose minimal plates

171. The non-overlapping nature of the genetic code is supported by the fact that

(A) poly(U-G) directs the synthesis of poly(cys-val)
(B) a single base mutation alters only one amino acid in the resulting protein
(C) most amino acids are coded by more than one triplet
(D) ribosomal bindings of amino-acyl-tRNA is stimulated by tri-nucleotides
(E) several terminator triplets have been discovered

172. The major implication of the *cis/trans* test for a gene in heterozygous diploid cells is that

(A) when two mutant sites lie in different genes, they show complementation to give a wild phenotype in only the *cis* arrangement
(B) when two mutant sites lie in different genes, they show complementation to give a wild phenotype in only the *trans* arrangement
(C) when two mutant sites lie in the same gene, they show complementation to give a wild phenotype in only the *cis* arrangement
(D) when two mutant sites lie in the same gene, they show complementation to give a wild phenotype in only the *trans* arrangement
(E) when two mutant sites lie in the same gene, they show complementation to give a wild phenotype in either the *cis* or *trans* arrangement

173. A potent inhibitor of protein synthesis that acts as an analog of aminoacyl-tRNA is

(A) mitomycin C
(B) streptomycin
(C) nalidixic acid
(D) rifampicin
(E) puromycin

174. Xeroderma pigmentosum is an inherited human skin disease which causes a variety of phenotypic changes in skin cells exposed to sunlight. The molecular basis of the disease appears to be

(A) a defect in an excision-repair system which removes thymine dimers from DNA
(B) the inability of the cells to synthesize carotenoid-type compounds
(C) the induction of a virulent provirus upon ultraviolet exposure
(D) the inactivation of temperature-sensitive transport enzymes in sunlight
(E) a rapid water loss caused by defects in the cell membrane permeability

175. F' factors are frequently lost from their bacterial hosts upon storage. An $F'lac^+$ factor is best maintained in

(A) a lac^-recA^- strain
(B) a lac^+recA^+ strain
(C) a lac^+recA^- strain
(D) a lac^-recA^+ strain
(E) an Hfr strain

176. The system of structural genes and regulator elements for the synthesis of various enzymes in a metabolic pathway is best described as

(A) a genome
(B) a muton
(C) a cistron
(D) an operon
(E) a codon

30

177. A complete cycle in protein synthesis, from a free amino acid to its incorporation into a peptide, requires

(A) one ATP
(B) one ATP and one GTP
(C) one ATP and two GTP's
(D) two ATP's
(E) two ATP's and one GTP

178. In studies of the mechanism of bacterial DNA replication, 5-bromo-uracil is often used as an analog of thymidine in order to

(A) cause specific frame-shift mutations for sequencing studies
(B) stop DNA synthesis at sites of thymidine incorporation
(C) provide a reactive group in the DNA for the preparation of DNA affinity supports
(D) synthesize a heavier DNA which can be identified by centrifugation
(E) create specific sites in the DNA for mild chemical cleavage

179. The point of integration of the prophage in a lysogenic bacteria may be determined by conjugation with a phage-sensitive recipient. The phenomenon is called

(A) gradient of transmission
(B) mapping by interrupted matings
(C) episome transfer
(D) phenotypic lag
(E) zygotic induction

180. Acridines, such as proflavine, induce which of the following classes of genetic mutation?

(A) Transversion
(B) Transition
(C) Frame-shift
(D) Chromosomal rearrangement
(E) Specific transition G:C A:T

181. The increased specificity of T4 phage synthesis over host cell functions is best explained by

(A) viral specific changes in the host RNA polymerase which increase the specificity of phage DNA transcription
(B) viral specific changes in the host ribosomes which increase the specificity of phage mRNA translation
(C) the synthesis of viral-coded nucleases specific for host DNA
(D) the synthesis of viral-coded proteases specific for host proteins
(E) repression of host DNA by viral regulators

182. Lysogenic bacteria are immune to superinfection because

(A) the prophage may be induced to enter the lytic cycle by excision from the bacterial chromosome
(B) the prophage is not inherited as part of the bacterial chromosome
(C) lysogeny and immunity result from the operation of the same control system which represses the expression of phage genes needed for lytic development
(D) bacterial phage receptors in the cell wall are inactivated by the process of lysogeny
(E) lysogenic bacteria actively synthesize phage-specific nucleases which degrade any phage DNA introduced into the bacteria in a virulent manner

183. The lac operon is an example of what type of control system?

(A) Negative control of repression
(B) Negative control of induction
(C) Positive control of repression
(D) Positive control of induction
(E) Super-repression

184. Positive control of repression is best described as a control system in which the operon

(A) only functions after an inducer protein, which can be inactivated by a co-repressor, switch it on
(B) only functions after an inducer protein, which is activated by an inducer, switches it on
(C) functions unless a repressor protein, which can be inactivated by an inducer, switches i off
(D) functions unless a repressor protein, which is activated by a co-repressor, switches it off
(E) functions unless it is switched off by a derepressed repressor protein

185. The function of a repressor protein in an operon system is to bind to

(A) an initiator formylmethionine (fMet)-tRNA, preventing protein synthesis
(B) the ribosome, preventing protein synthesis
(C) the RNA polymerase, preventing transcription
(D) a specific region of the mRNA, preventing translation to protein
(E) a specific region of the operon preventing the transcription of structural genes

186. Hot spots are

(A) DNA regions of low recombination frequency
(B) localized areas of energy buildup within the cell
(C) sites in proteins of high susceptibility to amino acid modification
(D) sites in DNA of high susceptibility to mutagenesis
(E) radioactive fragments of DNA formed during in vitro replication experiments

187. In the method of genetic transfer called transduction,

(A) the donor bacterium unites with a recipient bacterium and transfers to the recipient a part of its chromosome
(B) purified DNA from the cells of the donor strain is used to induce permanent hereditary changes in the cells of a second strain
(C) modified particles of bacterial virus of low virulence act as vectors of bacterial genes
(D) the sex factor of an F^+ strain is transferred to an F^- strain
(E) large chromosomal regions are transferred, thereby facilitating the mapping of genes in general

DIRECTIONS: Each question below contains four suggested answers of which **one** or **more** is correct. Choose the answer

A	if	**1, 2, and 3**	are correct
B	if	**1 and 3**	are correct
C	if	**2 and 4**	are correct
D	if	**4**	is correct
E	if	**1, 2, 3, and 4**	are correct

188. Oligomycin is thought to interfere with synthesis of energy-rich compounds, e.g., ATP, by

(1) dissociating cytochrome c from mitochondrial membranes
(2) uncoupling electron transfer between NAD and flavoprotein
(3) blocking the carnitine shunt in mitochondria
(4) inhibiting mitochondrial ATPase

189. Protein biosynthesis requires

(1) mRNA
(2) peptidyl transferase
(3) ribosomes
(4) ATP

SUMMARY OF DIRECTIONS

A	B	C	D	E
1, 2, 3 only	1, 3 only	2, 4 only	4 only	All are correct

190. Covalent bonds in DNA include

(1) adenine → β-glycosidic linkage to C-1′ of deoxyribose
(2) phosphate → 2′-OH of deoxyribose
(3) phosphate → 5′-OH of deoxyribose
(4) phosphate → phosphate

191. Inhibitors of nucleic acid biosynthesis include

(1) methotrexate
(2) chloramphenicol
(3) actinomycin D
(4) tetracycline

192. Donor cells of chromosomal DNA in an F^+ culture are called Hfr (high-frequency recombination) cells. Which of the following statements concerning them are true?

(1) Conjugation with an Hfr strain usually results in transfer of a complete chromosome
(2) Hfr cells transmit free F (sex factor) particles
(3) Hfr cells can not revert back to F^+ when the sex factor is detached from the chromosome
(4) Each Hfr cell has its own specific origin and order of gene transfer

193. Which of the following statements about conjugation are true?

(1) Physical contact between the donor and recipient cell is required
(2) It results in the transfer of only small fragments of the donor's chromosomes
(3) Cells lacking the sex factor (F^-) can act only as recipients
(4) Cells possessing the sex factor (F^+) can act only as donors

194. Which of the following statements concerning the genetic mechanisms of bacterial drug resistance are true?

(1) Mutations occur at a rate of 10^{-5} to 10^{-9} per cell, per generation
(2) Recombination between two cells, each resistant to a different drug, can produce a cell resistant to both
(3) A plasmid-carrying cell introduced into a cell population usually infects every other cell
(4) Staphylococci that produce penicillinase do not arise by chromosomal gene mutation

195. The "early proteins" of the T-even bacteriophages include

(1) enzymes for hydroxymethylcytosine production
(2) head and tail subunits
(3) DNA polymerase
(4) lysozyme for cell wall degradation

196. The sex factor in an Hfr cell is

(1) integrated in the chromosome
(2) associated with a particular sequence of gene transfer
(3) an endogenote
(4) autotransferable in the same manner as the F factor

197. Inhibition of bacterial cell wall synthesis is the primary metabolic effect of

(1) chloramphenicol
(2) bacitracin
(3) sulfonamide
(4) penicillin

198. Microbial cells possessing the sex factor (F) can

(1) act as males
(2) act as females
(3) conjugate with females
(4) conjugate with males

199. Chloramphenicol

(1) interferes with cell wall synthesis in bacteria
(2) is bacteriostatic for many bacteria
(3) interferes with RNA synthesis
(4) interferes with the incorporation of amino acids

200. Independent DNA replication occurs in

(1) bacteria
(2) mitochondria
(3) nuclei
(4) viruses

201. Which of the following compounds exert their antibiotic effect on bacterial ribosomes?

(1) Chloramphenicol
(2) Gentamycin
(3) Streptomycin
(4) Tetracycline

202. To label RNA radioactively one could use tritiated

(1) thymine
(2) adenine
(3) deoxyribose
(4) uracil

203. Which of the following statements are true concerning the reduction of purine ribonucleotides to purine deoxyribonucleotides?

(1) NADPH is the ultimate source of reducing equivalents
(2) It involves the same reductase system for all four nucleosides
(3) It takes place at the level of the ribonucleotide diphosphate
(4) It involves cleavage of the glycosidic bond in the sugar nucleotide

204. The nitrogens of the purine ring are derived from

(1) aspartic acid
(2) glycine
(3) glutamine
(4) serine

SUMMARY OF DIRECTIONS

A	B	C	D	E
1, 2, 3 only	1, 3 only	2, 4 only	4 only	All are correct

205. Which of the following statements are true about the regulation of the de novo synthesis of purine nucleotides?

(1) The rate-limiting step is thought to be the synthesis of 5-phosphoribosylamine from 5-phosphoribosylpyrophosphate and glutamine
(2) The rate-limiting step is subject to feedback inhibition by adenine nucleotides
(3) The rate-limiting step is subject to feedback inhibition by guanine nucleotides
(4) The rate-limiting step is subject to feedback inhibition by uric acid

206. Uric acid is a breakdown product of

(1) AMP
(2) GMP
(3) IMP
(4) CMP

207. Nucleoprotein complexes include

(1) viruses
(2) the translational complex
(3) ribosomes
(4) cell membranes

208. The thermal denaturation of DNA is characterized by

(1) a decrease in light absorption at 260 nm
(2) the formation of a triple helix
(3) a broad 10° C denaturation range for a homogenous DNA
(4) the melting point temperature varying proportionately to the guanine - cytosine base pair content

209. Which of the following statements concerning RNA and DNA polymerases are true?

(1) RNA polymerases require primers, and add bases at the 5′-end of the growing polynucleotide chain
(2) DNA polymerases can add nucleotides at both ends of the chain
(3) DNA polymerase can only synthesize DNA in the presence of an RNA template
(4) All RNA and DNA polymerases can only add nucleotides at the 3′-end of the growing polynucleotide chain

210. DNA-dependent RNA polymerase from *E. coli* may

(1) be a simpler molecule than *E. coli* DNA polymerase
(2) not require all four nucleotide precursors to initiate RNA synthesis
(3) synthesize RNA in a 3′ to 5′ direction
(4) not require sigma factor for chain elongation

211. Operons may contain

(1) single structural genes
(2) multiple structural genes
(3) nucleotides coding for a single messenger RNA
(4) repressor genes, regulator genes, and operator genes

212. Operator genes

(1) synthesize repressors
(2) affect structural genes
(3) synthesize aporepressors
(4) are affected by repression

213. Studies of the genetic code have revealed that

(1) the nucleotide on the 3'-end of a triplet has the least specificity for an amino acid
(2) no signal exists to indicate the end of one codon and the beginning of another
(3) only three triplets are "nonsense" triplets
(4) mRNA molecules specify only one polypeptide chain

214. Which of the following methods may be employed in "nearest neighbor" analysis of RNA?

(1) Synthesis of RNA with ^{14}C-labeled ribonucleotide triphosphates
(2) Alkaline hydrolysis of RNA products
(3) Hybridization to complementary DNA
(4) Synthesis of RNA with only one of the four ribonucleotide triphosphates labeled with ^{32}P at the alpha position

215. Which of the following are correct Watson-Crick base pairs in a DNA double helix?

(1) Adenine-thymine
(2) Uracil-adenine
(3) Cytosine-guanine
(4) Guanine-adenine

216. During translation

(1) three nucleotide bases code for one amino acid
(2) specific nucleotide sequences signal peptide chain termination
(3) the last nucleotide in a codon has less specificity than the others
(4) more than one group of nucleotides may code for a single amino acid

217. Cell fusion

(1) occurs between ova and sperm
(2) is part of conjugation in *E. coli*
(3) allows for genetic recombination
(4) is required for log phase growth

218. During the course of protein synthesis

(1) amino acids are attached randomly to tRNAs
(2) the nascent peptide chain is synthesized from its carboxyl terminus
(3) the mRNA is pulled along the ribosomes by contractile microtubules
(4) the growing peptide chain is attached to a tRNA

DIRECTIONS: The groups of questions in this section consist of four or five lettered headings followed by several numbered items. For each numbered item choose the **one** lettered heading with which it is **most** closely associated. Each lettered heading may be used once, more than once, or not at all.

Questions 219-225

For each disorder, select the mode of inheritance with which it is most likely to be associated.

(A) Autosomal dominant inheritance
(B) Autosomal recessive inheritance
(C) X-linked dominant inheritance
(D) X-linked recessive inheritance

219. Huntington's chorea

220. Vitamin D-resistant rickets

221. Duchenne's hypertrophic muscular dystrophy

222. Phenylketonuria

223. Cystic fibrosis

224. Red-green blindness

225. Wilson's disease

Questions 226-229

For each of the RNA species, choose the description with which it is most likely to be associated.

(A) Part of the ribosomal subunit with peptidyl transferase activity
(B) Part of the ribosomal subunit that binds mRNA
(C) Part of the chromosome
(D) Carries the anticodon and serves as adaptor to translate the genetic code into a sequence of amino acids
(E) Complementary to the base sequence of an operon and codes, therefore, for the polypeptide chains that correspond to the genes of the operon

226. mRNA

227. tRNA

228. 23 S rRNA

229. 16 S rRNA

Carbohydrates and Lipids

DIRECTIONS: Each question below contains five suggested answers. Choose the **one best** response to each question.

230. Which of the following enzymes function in both glycolysis and gluconeogenesis?

(A) Pyruvate kinase
(B) Pyruvate carboxylase
(C) Glyceraldehyde 3-phosphate dehydrogenase
(D) Fructose 1,6-diphosphatase
(E) Hexokinase

231. The structure shown below is that of

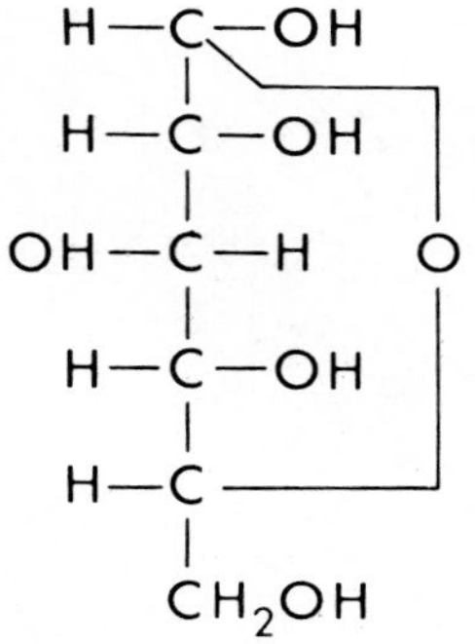

(A) α-D-glucopyranose
(B) β-D-glucopyranose
(C) α-D-glucofuranose
(D) β-L-glucofuranose
(E) α-D-fructofuranose

232. Gluconeogenesis is a process in carbohydrate metabolism which forms

(A) glucose
(B) maltose
(C) sucrose
(D) fructose
(E) glucose 1-phosphate

233. The formation of carbohydrates from amino acids is known as

(A) glycolysis
(B) glycogenolysis
(C) glucolysis
(D) gluconeogenesis
(E) glycogenesis

234. Hydrolysis of maltose will yield

(A) glucose only
(B) fructose + glucose
(C) galactose + glucose
(D) mannose + glucose
(E) fructose + galactose

235. The first product of glycogenolysis is

(A) glucose 6-phosphate
(B) glucose 1,6-diphosphate
(C) glucose 1-phosphate
(D) fructose 1-phosphate
(E) glucose

236. The substrate for aldolase is

(A) glucose 6-phosphate
(B) fructose 6-phosphate
(C) fructose 1,6-diphosphate
(D) phosphoglyceric acid
(E) 1,3-diphosphoglyceric acid

237. In the pathway below, ATP is produced between

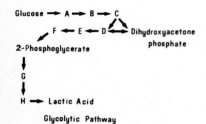

Glucose → A → B → C
F ← E ← D ← Dihydroxyacetone phosphate
2-Phosphoglycerate
G
H → Lactic Acid
Glycolytic Pathway

(A) A-B
(B) A-B and G-H
(C) B-C
(D) C-D
(E) G-H

238. Which of the following fatty acids is a biosynthetic precursor of prostaglandin $F_1 \alpha$?

(A) Oleic acid
(B) Stearic acid
(C) Palmitic acid
(D) Homo-γ-linolenic acid
(E) None of the above

239. An animal deficient in essential fatty acids would accumulate which of the following polyunsaturated fatty acids?

(A) Docosahexenoic acid
(B) Arachidonic acid
(C) Eicosa-11,14-dienoic acid
(D) γ-Linolenic acid
(E) Eicosa-5,8,11-trienoic acid

240. Which of the following fatty acids can be synthesized de novo in higher animals?

(A) Eicosa-5, 8, 11-trienoic acid
(B) Arachidonic acid
(C) Docosahexenoic acid
(D) Linolenic acid
(E) Linoleic acid

241. Diglyceride + NuDP-choline ⇒ NuMP + phosphatidylcholine. In the above reaction, NuMP stands for

(A) AMP
(B) CMP
(C) GMP
(D) TMP
(E) UMP

242. The cholesterol molecule is

(A) an aromatic ring
(B) a straight chain acid
(C) a steroid
(D) a tocopherol
(E) a quinoline derivative

243. Cholesterol is synthesized from

(A) acetyl CoA
(B) malate
(C) α-ketoglutaric acid
(D) oxalate
(E) pyruvate

244. Which step in the biosynthesis of cholesterol is thought to be rate-controlling and the locus of metabolic regulation?

(A) Geranyl pyrophosphate → farnesyl pyrophosphate
(B) Squalene → lanosterol
(C) Lanosterol → cholesterol
(D) 3-Hydroxy-3-methylglutaryl CoA → mevalonic acid
(E) None of the above

245. Which of the following sterols is formed directly by the cyclization of squalene in mammals?

(A) Cholesterol
(B) Desmosterol
(C) Lanosterol
(D) β-Sitosterol
(E) Cortisol

246. Sterols are metabolized to

(A) coenzyme A
(B) precursors of cholesterol
(C) vitamin A
(D) vitamin D
(E) vitamin E

247. A low serum carotene level is caused by

(A) vitamin A deficiency
(B) intrinsic factor deficiency
(C) malabsorption syndrome
(D) keratoconjunctivitis sicca
(E) achlorhydria

248. Which of the following is a phospholipid?

(A) Glycogen
(B) Sphingomyelin
(C) Prostaglandin
(D) Oleic acid
(E) Triglyceride

249. All of the following compounds are phospholipids EXCEPT

(A) cephalins
(B) cerebrosides
(C) lecithins
(D) sphingomyelins
(E) plasmalogens

250. The phospholipid cardiolipin is found almost exclusively in

(A) mitochondrial membranes
(B) plasma membranes
(C) lysosome membranes
(D) smooth endoplasmic reticular membranes
(E) rough endoplasmic reticular membranes

251. Which of the following is NOT a constituent of ganglioside molecules?

(A) Glycerol
(B) Sialic acid
(C) Hexose sugar
(D) Sphingosine
(E) Long-chain fatty acid

252. Which of the following compounds has the lowest density?

(A) Chylomicrons
(B) β-Lipoproteins
(C) Pre-β-lipoproteins
(D) α-Lipoproteins
(E) Transferrin

253. Cyanides produce hypoxia by

(A) producing central hypoventilation
(B) interfering with oxygen carriage
(C) slowing capillary circulation
(D) inhibiting cellular respiration
(E) none of the above mechanisms

254. The most appropriate term for the phosphorylation of glucose during its entry into a bacterial cell is

(A) active transport
(B) chemoperfusion
(C) facilitated diffusion
(D) group translocation
(E) passive transport

255. All of the following statements about blood-glucose levels in humans are true EXCEPT

(A) normal fasting values are usually between 70 and 110 mg %
(B) normal values may exceed diabetic values given different physiologic conditions (postprandial normal—fasting diabetic)
(C) renal thresholds for glucose are usually exceeded in fasting diabetics
(D) after ingestion of large amounts of sugar, blood-glucose levels may attain 200 mg % in normal patients
(E) Postprandial blood-sugar levels decrease from peak values more rapidly in normal individuals than in diabetics

256. What is the net ATP yield in converting one glucosyl residue in glycogen to two molecules of lactate?

(A) One
(B) Two
(C) Three
(D) Four
(E) Five

257. The ratio of the number of high-energy bonds formed per mole of glucose under aerobic conditions to the number formed per mole of glucose under anaerobic conditions is

(A) 2:1
(B) 9:1
(C) 13:1
(D) 19:1
(E) 25:1

258. Under aerobic conditions in the glycolytic pathway

(A) there is no net change in the oxidation-reduction state of NAD
(B) there is net consumption of ATP
(C) no oxidation-reduction steps take place
(D) glucose is degraded to ethanol
(E) citric acid is an intermediate compound

259. Fluoride is an inhibitor of which of the following enzymes of the glycolytic pathway?

(A) Hexokinase
(B) Aldolase
(C) Pyruvate kinase
(D) Enolase
(E) Phosphohexose isomerase

260. The net number of ATP molecules formed per molecule of glucose in aerobic glycolysis is

(A) 2
(B) 8
(C) 18
(D) 36
(E) 54

Diagram accompanies Question 261

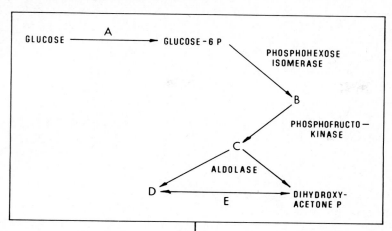

261. In the figure shown above fructose 1,6-diphosphate is located at point

(A) A
(B) B
(C) C
(D) D
(E) E

262. Which of the following energy-related reactions does NOT occur in mitochondria?

(A) Citric acid cycle
(B) Fatty acid oxidation
(C) Electron transport
(D) Glycolysis
(E) Oxidative phosphorylation

263. Which of the following compounds does NOT contain a high-energy phosphate bond?

(A) ADP
(B) Creatine phosphate
(C) Glucose 6-phosphate
(D) Phosphoenolpyruvate
(E) 1,3-Diphosphoglycerate

264. In the diagram of ADP shown below, four bonds are labeled A-D. Which is a high-energy bond?

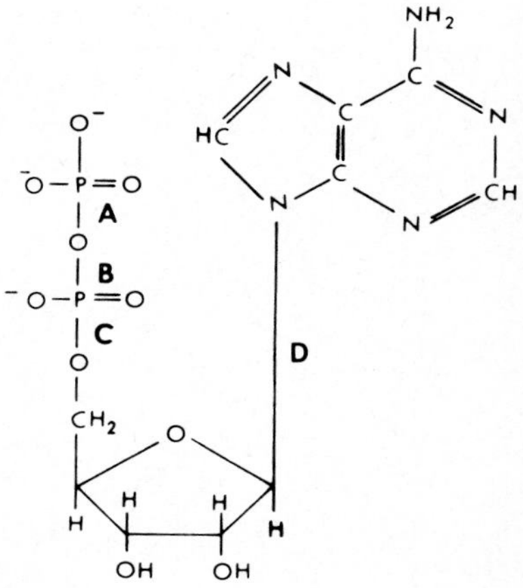

(A) A
(B) B
(C) C
(D) D
(E) None of the above

265. Among the many molecules of high-energy phosphate compounds which are by-products of the Krebs tricarboxylic acid cycle, one molecule is synthesized at the substrate level. In which of the following steps does this occur?

(A) Citrate → α-ketoglutarate
(B) α-Ketoglutarate → succinate
(C) Succinate → fumarate
(D) Fumarate → malate
(E) Malate → oxaloacetate

266. In glycolysis the following reactions occur with the following free energy changes:

Glyceraldehyde 3-phosphate + NAD^+ + P_i ⇌ 1,3-diphosphoglycerate + NADH + H^+: $\Delta F^{o'}$ = +1.5 Kcal/mole

1,3-Diphosphoglycerate + ADP ⇌ 3-phosphoglycerate + ATP: $\Delta F^{o'}$ = −4.5 Kcal/mole

For the two step process converting glyceraldehyde 3-phosphate to 3-phosphoglycerate, the overall free energy change is

(A) $\Delta F^{o'}$ = +3.0 Kcal/mole
(B) $\Delta F^{o'}$ = −3.0 Kcal/mole
(C) $\Delta F^{o'}$ = +6.0 Kcal/mole
(D) $\Delta F^{o'}$ = −6.0 Kcal/mole
(E) $\Delta F^{o'}$ = −4.5 Kcal/mole

267. What is the substrate for the only physiologically irreversible reaction in the Krebs citric acid cycle?

(A) Citrate
(B) α-Ketoglutarate
(C) Succinate
(D) Fumarate
(E) Malate

268. The diagram of the citric acid cycle shown below, contains lettered steps where H^+-e^- pairs might be given to the electron transport chain. At which step does donation of an H^+-e^- pair NOT occur?

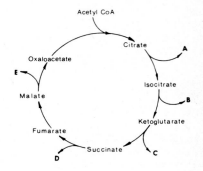

(A) A
(B) B
(C) C
(D) D
(E) E

269. Which of the following enzymes is NOT involved in the citric acid (Krebs) cycle?

(A) Fumarase
(B) Isocitrate dehydrogenase
(C) Succinate thiokinase
(D) Pyruvate dehydrogenase
(E) Aconitase

270. Which of the following sources of energy can NOT be used by living cells for their metabolic functions?

(A) ATP
(B) Fats
(C) Heat
(D) Sugars
(E) Sunlight

44

271. The citric acid cycle

(A) contains no intermediates for glucogenesis
(B) contains intermediates for amino acid synthesis
(C) generates fewer molecules of ATP than glycolysis, per mole of glucose consumed
(D) is an anaerobic process
(E) is the major anabolic pathway for glucose synthesis

272. An allosteric enzyme responsible for controlling the rate of the citric acid cycle is

(A) citrate dehydrogenase
(B) isocitrate dehydrogenase
(C) malate dehydrogenase
(D) aconitase
(E) pyruvate dehydrogenase

273. The monosaccharide most rapidly absorbed from the small intestines is

(A) xylose
(B) glucose
(C) fructose
(D) mannose
(E) galactose

274. Which of the following comparisons between hexokinase and glucokinase is FALSE?

(A) The Michaelis constant of hexokinase for glucose is much smaller than that of glucokinase
(B) Hexokinase is less specific about sugars it will accept as substrate than is glucokinase
(C) Only hexokinase is inhibited by glucose 6-phosphate
(D) Only glucokinase is present in the brain
(E) Hexokinase and glucokinase are present in liver tissue

275. What is the Cori cycle?

(A) The interconversion between glycogen and glucose 1-phosphate
(B) The synthesis of alanine from pyruvate in skeletal muscle and the synthesis of pyruvate from alanine in liver
(C) The synthesis of urea in liver and degradation of urea to carbon dioxide and ammonia by bacteria in the gut
(D) The production of lactate from glucose in peripheral tissues with the resynthesis of glucose from lactate in liver
(E) None of the above

276. Which of the following is NOT a disaccharide?

(A) Cellobiose
(B) Cellulose
(C) Lactose
(D) Maltose
(E) Sucrose

277. One of the principal sources of the hydrogen stored in the form of NADPH is

(A) synthesis of fatty acids
(B) oxidative phosphorylation
(C) glycolysis
(D) the citric acid cycle
(E) the hexose monophosphate shunt (phosphogluconate pathway)

278. How many ATP molecules are required to convert two molecules of lactate into glucose in mammalian liver?

(A) Two
(B) Three
(C) Four
(D) Five
(E) Six

279. Gluconeogenesis refers to the intracellular synthesis of glucose from other metabolic precursors. Which of the following compounds will NOT yield a net synthesis of glucose?

(A) Aspartic acid
(B) Glutamic acid
(C) Leucine
(D) Phosphoenolpyruvate
(E) Succinic acid

280. In the reaction below, NuDP stands for

NuDP glucose + glycogen$_n$ $\rightarrow$ NuDP + glycogen$_{n+1}$

(A) ADP
(B) CDP
(C) GDP
(D) TDP
(E) UDP

281. In the reaction below, NuTP stands for

NuTP + glucose $\rightarrow$ glucose 6-phosphate + NuDP

(A) ATP
(B) CTP
(C) GTP
(D) TTP
(E) UTP

282. In the reaction below, NuTP stands for

oxaloacetate + NuTP $\rightarrow$ NuDP + phosphoenolpyruvate + CO_2

(A) ATP
(B) CTP
(C) GTP
(D) TTP
(E) UTP

Questions 283-284

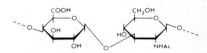

283. The structure shown above is the mucopolysaccharide

(A) chitin
(B) chondroitin sulfate
(C) heparin
(D) hyaluronic acid
(E) keratin

284. The mucopolysaccharide shown above is found primarily in the

(A) cartilage
(B) hard, outer shells of insects
(C) mast cells
(D) outer layers of the epidermis
(E) vitreous humor of the eye

285. Which is the active form of sulfate used in the biosynthesis of sulfate ester?

(A) Chondroitin sulfate
(B) 3'-Phosphoadenosine 5'-phosphosulfate
(C) Adenosine 5'-phosphosulfate
(D) Sulfonyl urea
(E) None of the above

286. Hydrolysis of a fat by an alkali is given the special name of

(A) esterification
(B) reduction
(C) saponification
(D) oxidation
(E) hydrolysis

287. The de novo biosynthesis of fatty acids

(A) does not utilize acetyl CoA
(B) produces only fatty acids shorter than ten carbon atoms
(C) requires the intermediate malonyl CoA
(D) takes place primarily in mitochondria
(E) uses NAD as an oxidizing agent

288. In the pathway leading to biosynthesis of acetoacetate from acetyl CoA in liver, which compound is the immediate precursor of acetoacetate?

(A) 3-Hydroxybutyrate
(B) Acetoacetyl CoA
(C) 3-Hydroxybutyryl CoA
(D) Mevalonic acid
(E) 3-Hydroxy-3-methylglutaryl CoA

289. The rate-limiting step in the extramitochondrial synthesis of fatty acids is

(A) condensing enzyme
(B) hydratase
(C) acetyl CoA carboxylase
(D) acyl transferase
(E) palmitoyl deacylase

290. Continuation of the fatty acid oxidation cycle in the metabolism of long-chain fatty acids is dependent on the presence of all of the following enzymes EXCEPT

(A) acyl CoA dehydrogenase
(B) β-hydroxyacyl CoA dehydrogenase
(C) enoyl hydrase
(D) β-ketothiolase
(E) thiokinase

291. In the biosynthesis of fatty acids, which is the compound that transports the acetate groups out of the mitochondria into the cytoplasm?

(A) Acetyl CoA
(B) Acetyl carnitine
(C) Citrate
(D) Acetyl phosphate
(E) None of the above

292. The oxidation and degradation of fatty acids in the cell

(A) begins with the fatty acid thioester of CoA
(B) does not produce useful energy for the cell
(C) is inhibited by carnitine
(D) occurs primarily in the nucleus
(E) proceeds through successive shortening of the fatty acids by three-carbon units

293. The enzyme which catalyzes the ATP-requiring conversion of fatty acids to an activated form, in the synthesis of triglycerides, is

(A) condensing enzyme
(B) thiolase
(C) thiokinase
(D) lipase
(E) glycerokinase

DIRECTIONS: Each question below contains four suggested answers of which **one** or **more** is correct. Choose the answer

A	if	1, 2, and 3	are correct
B	if	1 and 3	are correct
C	if	2 and 4	are correct
D	if	4	is correct
E	if	1, 2, 3, and 4	are correct

294. Which of the following sugars would be expected to form the same osazones as glucose?

(1) Mannose
(2) Galactose
(3) Fructose
(4) Ribose

295. Which of the following sugars are ketose sugars?

(1) Ribose
(2) Ribulose
(3) Glucose
(4) Fructose

296. Which of the following compounds contain sugar residues?

(1) ATP
(2) NAD
(3) RNA
(4) Acetyl CoA

297. Which of the following are substrates for gluconeogenesis in mammalian liver?

(1) Oleic acid
(2) Serine
(3) Leucine
(4) Glycerol

298. The polysaccharide glycogen is

(1) a structural support for liver cell membranes
(2) a copolymer of glucose and galactose
(3) branched at few points
(4) hydrolyzed by β-amylase to form the disaccharide maltose

299. Which of the following statements are true of the structure of glycogen?

(1) There are α-1,4 glycosidic linkages
(2) There are α-1,6 glycosidic linkages
(3) All of the monosaccharides in glycogen are α-D-glucose
(4) Glycogen is an unbranched molecule

300. The irreversible enzymes bypassed in gluconeogenesis include

(1) phosphofructokinase
(2) hexokinase
(3) pyruvate kinase
(4) enolase

48

301. Glycogen synthetase, the enzyme involved in the biosynthesis of glycogen, may

(1) be more specifically defined as a UDP-glucose-glycogen glycosyl transferase
(2) be able to synthesize glycogen without a polymer primer
(3) exist in active and inactive forms, subject in part to hormonal control of activation
(4) employ UDP-D-glucose as a glycosyl donor in both plants and animals

302. The hexose monophosphate shunt includes which of the following enzymes?

(1) Fumarase
(2) α-Ketoglutarate dehydrogenase
(3) Hexokinase
(4) Glucose 6-phosphate dehydrogenase

303. Pyruvate is an intermediate in, or an immediate precursor to, which of the following biosynthetic pathways?

(1) The citric acid cycle
(2) Lysine production in some bacteria
(3) Glycolysis
(4) Carbohydrate production

304. Oxidative decarboxylation of pyruvate involves

(1) NAD
(2) NADP
(3) FAD
(4) folic acid

305. Which of the following inhibit the citric acid cycle?

(1) Arsenite
(2) Malonate
(3) Fluoroacetate
(4) Anaerobic conditions

306. Factors affecting the activity of the citric acid cycle probably include

(1) levels of oxaloacetic acid
(2) levels of NAD^+
(3) ratio of concentrations of ADP/ATP
(4) number of mitochondria per cell

307. The reaction catalyzed by α-ketoglutarate dehydrogenase in the Krebs cycle requires

(1) NAD
(2) NADP
(3) CoA
(4) ATP

308. Glycolysis in the Embden-Meyerhof pathway is typified by

(1) the conversion of glucose to lactic acid in anaerobic mammalian muscle
(2) a requirement for O_2, in order for yeast to convert glucose to CO_2 and ethanol
(3) the ability to proceed independent of P_{O_2}
(4) a net gain of one mole of ATP per mole of glucose traversing the pathway under aerobic conditions

309. Which of the following are inhibitors of the enzyme phosphofructokinase?

(1) Citrate
(2) Cyclic AMP
(3) ATP
(4) NH_4^+

310. Which of the following statements are consistent with the chemosmotic hypothesis of Mitchell (a currently popular model to explain mitochondrial oxidative phosphorylation)?

(1) The energy for ATP synthesis is derived from re-entry of protons into the mitochondrion down an electrochemical gradient
(2) Uncouplers (dinitrophenol) can increase the permeability of artificial lipid membranes to protons
(3) As electrons are passed down the electron transport chain, there is a separation of charge which results in active transport out of mitochondria
(4) The pH is ordinarily lower inside the mitochondrion than outside

311. Which of the following statements describe the mechanism by which NADH is thought to be transported into mitochondria from the cytoplasm for the purpose of aerobic oxidation?

(1) NADH is transported across the mitochondrial membranes directly
(2) Dihydroxyacetone phosphate is reduced to glycerol 3-phosphate which is in turn oxidized by a flavoprotein in the inner membrane of the mitochondrion
(3) Oxaloacetate is reduced to malate which enters the mitochondrion where it is oxidized to oxaloacetate which then exits from the mitochondrion
(4) Oxaloacetate is reduced to malate which enters the mitochondrion where it is oxidized to oxaloacetate and then transaminated to aspartate for transport out of the mitochondrion

312. Which of the following statements about skeletal muscle glycogen phosphorylase are correct?

(1) Cyclic AMP leads indirectly to the conversion of glycogen phosphorylase to the active, phosphorylated form
(2) Glycogen phosphorylase is the enzyme responsible for both biosynthesis and degradation of glycogen
(3) AMP is an allosteric activator of the inactive form of the enzyme phosphorylase *b*
(4) Exposure of muscle to the hormone insulin leads to activation of glycogen phosphorylase

313. Which of the following statements concerning "energy-rich" compounds of biologic systems are true?

(1) Only phosphate esters serve as "energy-rich" compounds
(2) Amino acid esters have free energies of hydrolysis comparable to that of ATP
(3) ATP, an "energy-rich" compound, has a positive free energy of hydrolysis
(4) Resonance theory is the accepted mechanism to explain the nature of "energy-rich" compounds

314. High-energy phosphate bonds are found in

(1) phosphoenolpyruvate
(2) ATP
(3) creatine phosphate
(4) ADP

315. In the electron transport chain

(1) continuation of the process is dependent on oxidative phosphorylation
(2) three moles of ATP are formed during the transfer of electrons from NAD to oxygen
(3) free energy change is positive for the transfer of electrons from NADH to oxygen
(4) the iron ion in the porphyrin ring undergoes $Fe^{++} \rightleftharpoons Fe^{+++}$ transitions

316. Which of the following are examples of biologically important mucopolysaccharides?

(1) Neuraminidase
(2) Hyaluronic acid
(3) Cellulose
(4) Heparin

317. Polysaccharides

(1) provide a major biologic energy source
(2) may be found in either linear or branched forms
(3) are important structural elements in bacterial cell walls
(4) are informational molecules

318. Compounds which are normally used by the body to conjugate bile acids include

(1) glycine
(2) glucuronic acid
(3) taurine
(4) fatty acids

319. Lipids are

(1) an intracellular energy source
(2) poorly soluble in water
(3) structural components of membranes
(4) composed of only carbon, hydrogen, and oxygen

320. In bacteria with cell walls, which of the following saccharide residues are constituents of the peptidoglycan?

(1) Lactose
(2) Glucosamine
(3) Fructose
(4) Muramic acid

321. Ketone bodies

(1) include acetone and acetoacetic acid
(2) may be excreted in urine
(3) may result from starvation
(4) are related to diabetes insipidus

51

SUMMARY OF DIRECTIONS

A	B	C	D	E
1, 2, 3 only	1, 3 only	2, 4 only	4 only	All are correct

322. Important precursors in the synthesis of fatty acids in animal tissues include

(1) carnitine
(2) pyruvate
(3) ATP
(4) acetyl CoA

323. Which of the following fatty acids are essential for life in man?

(1) Palmitic acid
(2) Stearic acid
(3) Oleic acid
(4) Linolenic acid

324. Which of the following statements are true of the synthesis of fatty acids from acetyl CoA but not of the oxidation of fatty acids to acetyl CoA?

(1) All oxidation-reduction steps use NADPH as a cofactor
(2) CoA is the only pantetheine-containing substance involved in the pathway
(3) Malonyl CoA is an "activated" intermediate
(4) The reactions proceed within mitochondria

325. Which of the following are intermediates in the metabolism of propionic acid?

(1) Propionyl CoA
(2) S-Methylmalonyl CoA
(3) R-Methylmalonyl CoA
(4) Succinyl CoA

326. Ethanol is converted in the liver to

(1) acetone
(2) acetaldehyde
(3) methanol
(4) acetyl CoA

327. Biosynthetic precursors of cholesterol include

(1) lanosterol
(2) mevalonic acid
(3) squalene
(4) progesterone

328. An 11β-hydroxylase is required in the biosynthesis of

(1) estradiol
(2) cortisol
(3) testosterone
(4) aldosterone

329. A 17-hydroxylase is required in the biosynthesis of

(1) estradiol
(2) cortisol
(3) testosterone
(4) aldosterone

330. A 21-hydroxylase enzyme is involved in the biosynthesis of

(1) estradiol
(2) cortisol
(3) testosterone
(4) aldosterone

331. Hydrolysis of a mixture of phosphoglycerides will yield

(1) choline
(2) glycerol
(3) phosphate
(4) serine

332. Chylomicrons are composed of

(1) triglycerides
(2) cholesterol
(3) phospholipids
(4) protein

333. The glyoxylate cycle

(1) is found in microorganisms as well as in higher forms of life
(2) oxidizes acetate completely to CO_2
(3) functions as an alternate to the Krebs cycle in bacteria
(4) is essential for the growth of microorganisms in media which contain only acetate as a carbon source

334. The presence of starch granules indicates

(1) an animal cell
(2) activation of the Calvin cycle ("dark reaction")
(3) that a prokaryote has metabolized glucose
(4) activation of the pentose phosphate pathway

335. Levan is

(1) a polymer of D-fructose
(2) analogous to dextran
(3) formed from sucrose by levansucrase
(4) a secondary metabolite of *Bacillus brevis*

336. Dextran is

(1) a long-chain polymer of sucrose
(2) used as a substitute for plasma in cases of blood loss
(3) a long-chain fructose polymer
(4) a long-chain glucose polymer

337. Lipid A

(1) forms part of the plasma membrane
(2) is part of the ketodeoxyoctonic acid complex in the cell wall
(3) contains the core sugars
(4) is attached to the core polysaccharide

338. The beta-oxidation of fatty acids occurs mainly in the

(1) cytoplasm
(2) cell membrane
(3) absence of ATP
(4) mitochondria

339. Lipoic acid is essential for

(1) glycolysis
(2) acetyl CoA carboxylase reactions
(3) the transamination of ketoglutaric acid
(4) oxidative decarboxylation of pyruvate

340. Which of the following statements concerning the regulatory effects of citrate are correct?

(1) It activates phosphofructokinase
(2) It activates acetyl CoA carboxylase
(3) It activates enolase
(4) It acts similarly to isocitrate

Questions 341-343

11-KETOETIOCHOLANOLONE   ANDROSTERONE   CORTOL   TETRAHYDROCORTISOL

341. Which of the following compounds give a positive Porter-Silber color reaction, i.e., are "17-hydroxysteroids?"

(1) 11-Ketoetiocholanolone
(2) Cortol
(3) Androsterone
(4) Tetrahydrocortisol

342. Which of the following are 17-ketogenic steroids?

(1) 11-Ketoetiocholanolone
(2) Cortol
(3) Androsterone
(4) Tetrahydrocortisol

343. Which of the following are 17-ketosteroids?

(1) 11-Ketoetiocholanolone
(2) Cortol
(3) Androsterone
(4) Tetrahydrocortisol

SUMMARY OF DIRECTIONS

A	B	C	D	E
1, 2, 3 only	1, 3 only	2, 4 only	4 only	All are correct

344. In which subcellular fractions does the desaturation of essential fatty acids occur in animals?

(1) Nuclear
(2) Supernatants of 100,000 x g (60 min)
(3) Mitochondrial
(4) Microsomal

Questions 345-347
For each disaccharide, choose the monosaccharides of which it is composed.

(A) Fructose only
(B) Glucose only
(C) Glucose and fructose
(D) Glucose and galactose
(E) Fructose and galactose

345. Sucrose

346. Lactose

347. Maltose

Questions 348-349
For each description, choose the structural carbohydrate with which it is most likely to be associated.

(A) Keratan sulfate
(B) Teichoic acid
(C) Starch
(D) Cellulose
(E) Murein

348. Polysaccharide component of bacterial cell walls

349. Principal, final polysaccharide component of connective tissue

Questions 350-352
For each reaction, select the lettered glycolytic step with which it is coupled in the diagram below.

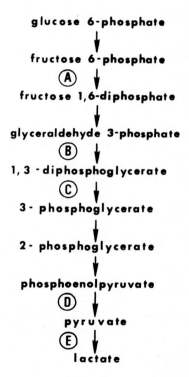

350. ATP → ADP

351. NADH → NAD

352. NAD → NADH

Questions 353-355

For each structure accompanied by its characteristics, choose the compound which it is most likely to be.

- (A) Prostaglandin A
- (B) Prostaglandin E
- (C) Prostaglandin F
- (D) Thyroid-stimulating hormone
- (E) Follicle-stimulating hormone

STRUCTURE	EFFECT ON BLOOD PRESSURE	NONVASCULAR SMOOTH MUSCLE STIMULATING ACTIVITY
353.	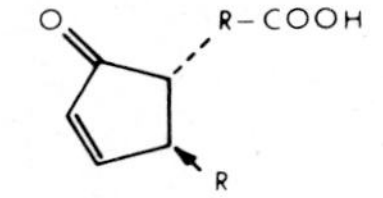 DECREASE	INACTIVE
354.	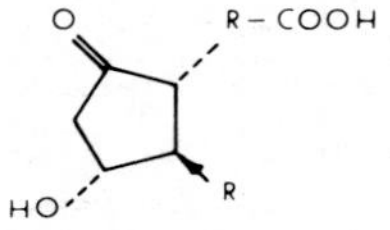 DECREASE	VERY ACTIVE
355.	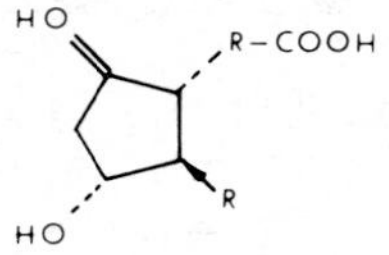 TRANSIENT INCREASE	VERY ACTIVE

Vitamins
and Hormones

DIRECTIONS: Each question below contains five suggested answers. Choose the **one best** response to each question.

356. A pellagra-like skin rash may be seen in

(A) phenylketonuria
(B) homogentisic aciduria
(C) Hurler's syndrome
(D) gargoylism
(E) Hartnup's disease

357. The earliest indication of lead intoxication is most likely to be

(A) increased excretion of delta-aminolevulinic acid in the urine
(B) punctate basophilia
(C) increased lead concentration in the blood
(D) increased coproporphyrin excretion in the urine
(E) free erythrocyte proto-porphyrin in excess of normal

358. The metal cofactor of tyrosinase is

(A) cobalt
(B) copper
(C) iron
(D) manganese
(E) magnesium

359. Which of the following hormones or compounds does NOT cause decreased lipolysis?

(A) Insulin
(B) Prostaglandin E
(C) Nicotinic acid
(D) Beta-adrenergic blockers
(E) Glucagon

360. In prolonged fasting, a limiting factor in the amount of glucose produced by the liver is

(A) guanidine
(B) alanine
(C) tyrosine
(D) tryptophan
(E) cytosine

361. Pairs of hormones with antagonistic effects include all of the following EXCEPT

(A) vasopressin − oxytocin
(B) insulin − glucagon
(C) melanocyte-stimulating hormone − melatonin
(D) calcitonin − parathyroid hormone
(E) histamine − serotonin

362. Insulin has many direct effects on various cell types from tissues such as muscle, adipose, liver, skin, etc. Which of the following cellular activities is NOT increased following exposure to physiologic concentrations of insulin?

(A) Plasma membrane transfer of glucose
(B) Glucose oxidation
(C) Gluconeogenesis
(D) Lipogenesis
(E) Formation of ATP, DNA, and RNA

363. Which of the following cellular activities is NOT decreased following exposure to physiologic concentrations of insulin?

(A) Ureogenesis
(B) Proteolysis
(C) Ketogenesis
(D) Glycogenolysis
(E) Proteogenesis

364. Evidence indicates that oxytocin and vasopressin are synthesized in the hypothalamic nuclei. They are then transported to their site of release in the posterior pituitary, in granular form, complexed with a large carrier protein called

(A) thymine
(B) transcortin
(C) angiotensin
(D) neurophysin
(E) cholecystokinin

365. Oxytocin, originally assayed by its ability to cause uterine muscle to contract, literally means "to stimulate birth." It is now known that oxytocin plays only a secondary role in uterine labor and that its major role is to

(A) increase blood pressure
(B) decrease blood pressure
(C) cause the placenta to release from the uterine wall
(D) cause the release of milk in the lactating breast
(E) control capillary permeability to oxygen in the lungs

366. Which of the following statements about prostaglandins is NOT true?

(A) They are cyclic fatty acids
(B) They have potent biologic effects that involve almost every organ in the body
(C) The first observation of their action was in causing uterine muscle to contract and blood pressure decrease
(D) Although they are found in many organs, they are only synthesized in the prostate gland and the seminal vesicles
(E) Medullin, isolated from the kidney medulla, is a prostaglandin

367. Insulin secretion in normal humans during constant intravenous glucose administration is best characterized by which of the following curves? (Glucose administration starts at time = 0.)

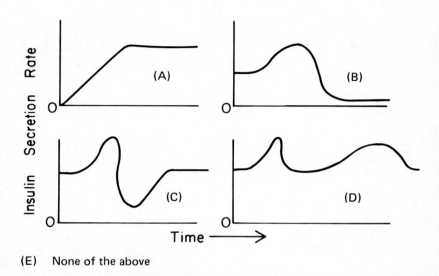

(E)   None of the above

368. Corneal vascularization may be a sign of deficiency of

(A)   thiamine
(B)   riboflavin
(C)   niacin
(D)   vitamin A
(E)   inositol

369. A patient's glucose tolerance test is shown below. The most likely diagnosis is

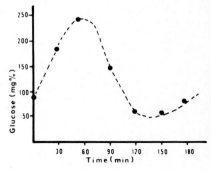

(A)   severe diabetes
(B)   Addison's disease
(C)   normal insulin response
(D)   delayed insulin response
(E)   von Gierke's disease

370. The hormones that differ the LEAST in their chemical structure are

(A) insulin and proinsulin
(B) oxytocin and vasopressin (ADH)
(C) cortisone and growth-stimulating hormone
(D) thyrotropin and oxytocin
(E) insulin and cortisone

Questions 371-372

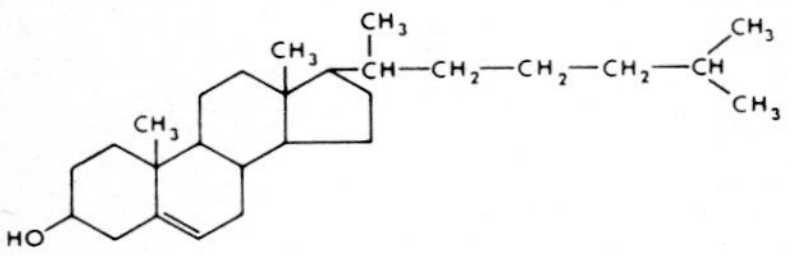

371. The structure shown above is a

(A) hormone
(B) cerebroside
(C) steroid
(D) terpene
(E) vitamin

372. The central ring structure shown above is common to all of the following compounds EXCEPT

(A) adrenocorticotropin
(B) aldosterone
(C) bile acids
(D) testosterone
(E) vitamin D

373. Which of the following hormones is not elaborated by the anterior pituitary in mammals?

(A) Somatotropin
(B) Prolactin
(C) Luteinizing hormone
(D) Vasopressin
(E) Follicle-stimulating hormone

374. Molecular iron, Fe, is

(A) stored primarily in the spleen
(B) excreted in the urine as Fe^{++}
(C) stored in the body in combination with ferritin
(D) absorbed in the intestine by transferrin
(E) absorbed in the ferric, Fe^{+3}, form

375. A lack of which of the following nutrients is best tolerated by humans?

(A) Protein
(B) Iodine
(C) Carbohydrate
(D) Lipid
(E) Calcium

376. Which of the following compounds is NOT a member of the electron transport chain?

(A) Ubiquinone (coenzyme Q)
(B) Cytochrome c
(C) NAD
(D) FAD
(E) Carnitine

377. A deficiency of vitamin B_{12} causes

(A) cheilosis
(B) beriberi
(C) pernicious anemia
(D) scurvy
(E) rickets

378. A severe deficiency of vitamin D in adults

(A) has no major effect
(B) causes rickets
(C) causes skin cancer
(D) causes osteomalacia
(E) causes night blindness

379. Classical scurvy results from a deficiency of

(A) thiamine
(B) riboflavin
(C) pantothenic acid
(D) ascorbic acid
(E) vitamin A

380. The coenzyme required in oxidative decarboxylation is

(A) biotin
(B) vitamin B_{12}
(C) pyridoxal phosphate
(D) ascorbic acid
(E) thiamine diphosphate

381. Pellagra can be prevented by treatment with

(A) thiamine
(B) niacin
(C) pyridoxine
(D) vitamin B_{12}
(E) pantothenic acid

382. Blockage of the absorption of fat-soluble vitamins would cause

(A) pernicious anemia
(B) scurvy
(C) pellagra
(D) rickets
(E) beriberi

383. Both Wernicke's disease and beriberi can be reversed by administering

(A) vitamin C
(B) vitamin A
(C) thiamine (B_1)
(D) vitamin B_6
(E) vitamin B_{12}

384. The coenzyme that functions with thiamine in the decarboxylation of α-keto acids to yield acyl CoA compounds is

(A) biotin
(B) lipoic acid
(C) vitamin A
(D) vitamin C
(E) NADP

385. The action of which of the following water-soluble vitamins is antagonized by methotrexate?

(A) Vitamin B_{12}
(B) Riboflavin
(C) Vitamin C
(D) Vitamin B_6 (pyridoxine)
(E) Folic acid

386. Pantothenic acid is a constituent of the coenzyme involved in

(A) decarboxylation
(B) acetylation
(C) dehydrogenation
(D) reduction
(E) oxidation

387. The compound shown below is important

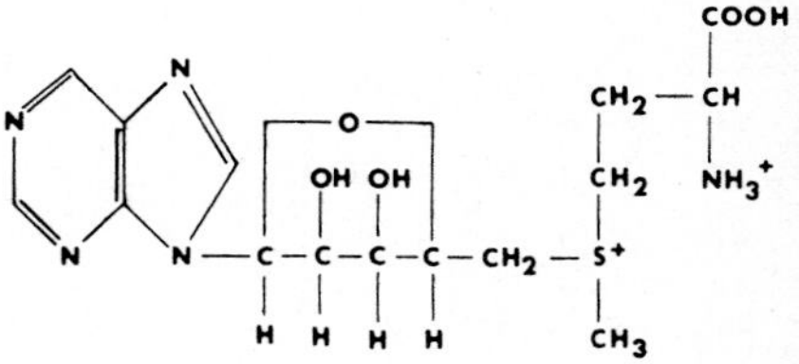

(A) in the formation of ATP
(B) in oxidative decarboxylation
(C) as a methyl donor
(D) as a gene repressor
(E) as a cofactor in transaminations

61

388. Which of the following cofactors is NOT involved in the pyruvate dehydrogenase reaction?

(A) Pyridoxal phosphate
(B) Thiamine pyrophosphate
(C) Lipoic acid
(D) FAD
(E) CoA

389. A deficiency of which of the following vitamins is specifically induced in animals by the administration of raw egg white?

(A) Pantothenate
(B) Riboflavin
(C) Thiamine
(D) Ascorbate
(E) Biotin

390. Biotin is involved in which of the following types of reactions?

(A) Decarboxylations
(B) Dehydrations
(C) Carboxylations
(D) Deaminations
(E) Hydroxylations

391. The administration of avidin, a biotin antagonist, would be expected to affect reactions catalyzed by all of the following enzymes EXCEPT

(A) succinic thiokinase
(B) propionyl CoA carboxylase
(C) β-methylcrotonyl CoA carboxylase
(D) acetyl CoA carboxylase
(E) pyruvic acid carboxylase

392. Which of the following vitamins is required for the action of transaminases?

(A) Niacin
(B) Pantothenate
(C) Thiamine
(D) Pyridoxal phosphate
(E) Riboflavin

393. Which of the following compounds is synthesized from glutamic acid, para-aminobenzoic acid, and a pteridine nucleus?

(A) Vitamin B_{12}
(B) Cyanocobalamin
(C) Folic acid
(D) Biotin
(E) CoA

394. Which of the following vitamins is the precursor of coenzyme A?

(A) Riboflavin
(B) Pantothenate
(C) Thiamine
(D) Cobamide
(E) Pyridoxamine

395. Vitamin E is a

(A) fatty acid
(B) propylthiouracil analog
(C) tocopherol
(D) quinone
(E) prostaglandin

Questions 396-397

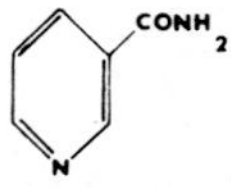

A

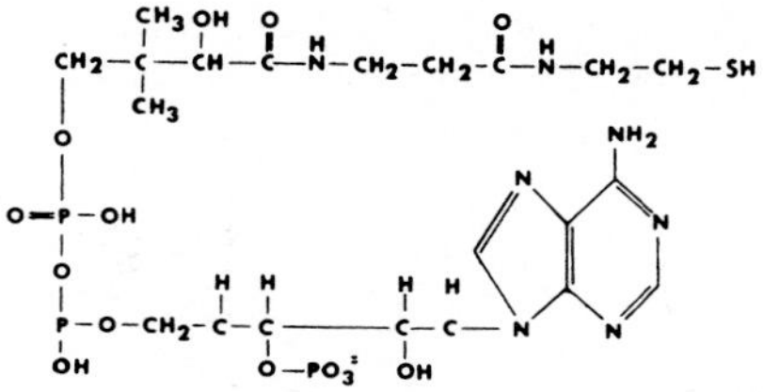

B

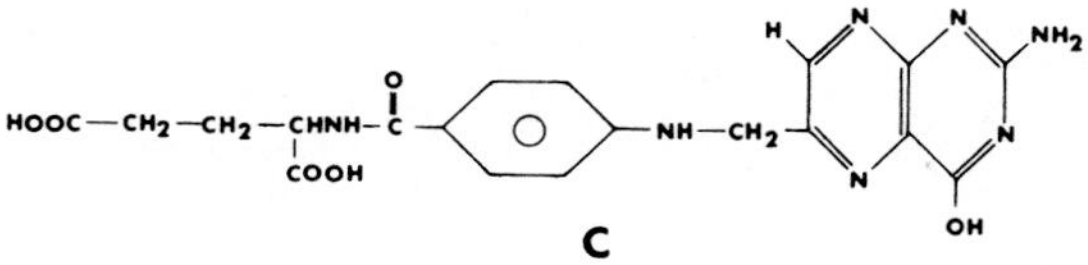

C

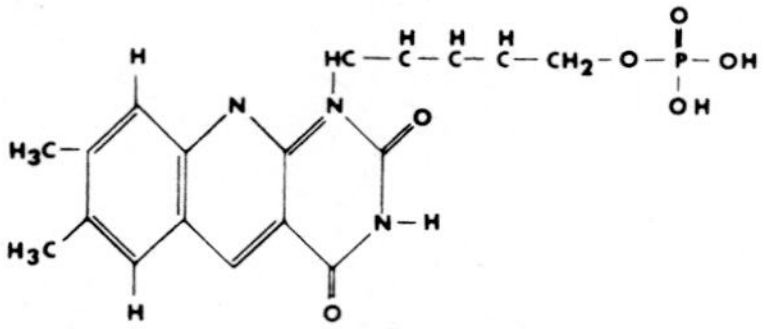

D

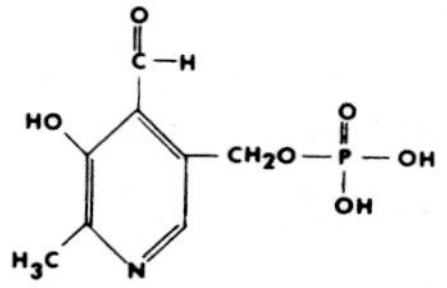

E

396. Which of the structures shown above is specific for the treatment of acute pellagra?

(A) A
(B) B
(C) C
(D) D
(E) E

397. A deficiency of compound **D** in the figure above can cause

(A) convulsive seizures
(B) liver damage
(C) pernicious anemia
(D) beriberi
(E) cheilosis

DIRECTIONS: Each question below contains four suggested answers of which **one** or **more** is correct. Choose the answer

A	if	**1, 2, and 3**	are	correct
B	if	**1 and 3**	are	correct
C	if	**2 and 4**	are	correct
D	if	**4**	is	correct
E	if	**1, 2, 3, and 4**	are	correct

398. Glucagon's effects include

(1) activating phosphorylase kinase
(2) inhibiting gluconeogenesis
(3) causing glycogenolysis
(4) activating muscle phosphorylase

399. Which of the following hormones are derived from the same amino acid?

(1) Epinephrine
(2) Vasopressin
(3) Thyroxine
(4) Prostaglandin E

400. Which of the following statements about the insulin molecule are true?

(1) After denaturation, it does not spontaneously renature to produce fully active molecules in good yield
(2) It is assembled from two fragmentary polypeptide chains
(3) It consists of two different polypeptide chains linked together by disulfide bonds
(4) Its subunit polypeptide chains are each biologically active with one-half the potency of the intact hormone

401. Which of the following hormones share one peptide chain in common with human chorionic gonadotropin (HCG)?

(1) LH
(2) TSH
(3) FSH
(4) PGH (human placental lactogen)

402. Coenzymes derived from the vitamin shown below are required by which of the following enzymes?

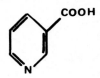

(1) Lactate dehydrogenase
(2) UDP glucose epimerase
(3) Phenylalanine hydroxylase
(4) Polynucleotide ligase of *E. coli*

403. Coenzymes derived from the vitamin shown below are required by which of the following enzymes?

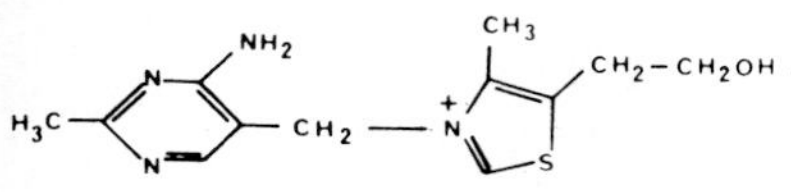

(1) Pyruvate dehydrogenase
(2) Transketolase (of pentose cycle)
(3) α-Ketoglutarate dehydrogenase
(4) Glutamate dehydrogenase

404. Coenzymes derived from the vitamin shown below are required by which of the following enzymes?

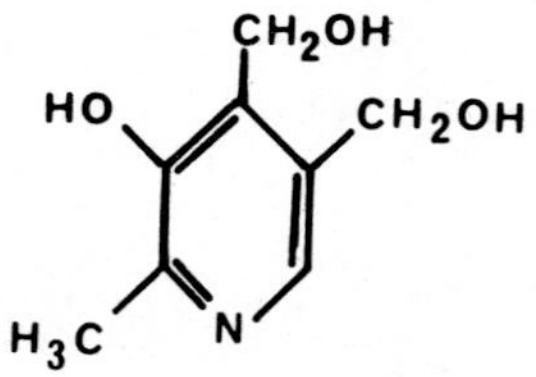

(1) Glutamic-oxaloacetic transaminase
(2) Glycogen phosphorylase
(3) Dihydroxyphenylalanine decarboxylase
(4) Aldolase

405. Coenzymes derived from the vitamin shown below are required by which of the following enzymes?

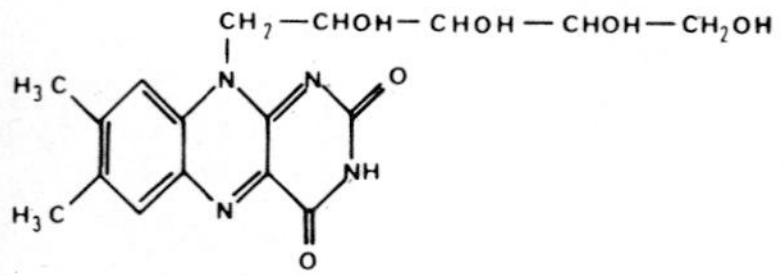

(1) Glutamate dehydrogenase
(2) Pyruvate dehydrogenase
(3) Lactate dehydrogenase
(4) Succinate dehydrogenase

406. Thiamine is important as a coenzyme for

(1) transketolases
(2) oxidative decarboxylases
(3) α-keto acid decarboxylases
(4) transaminases

407. Fat-soluble vitamins include

(1) cis-retinene
(2) calciferol
(3) tocopherol
(4) ascorbic acid

408. Vitamin B_6 (pyridoxine) is frequently a cofactor in

(1) decarboxylation
(2) deamination
(3) transamination
(4) transsulfuration

409. Hormones secreted by the pituitary include

(1) ACTH
(2) aldosterone
(3) FSH
(4) parathormone

410. Which of the following are water-soluble vitamins?

(1) Vitamin D
(2) Vitamin K
(3) Vitamin A
(4) Vitamin B_{12}

411. NAD, NADP, CoA, FAD, and ATP all contain

(1) at least one phosphate
(2) adenosine
(3) ribose
(4) a vitamin

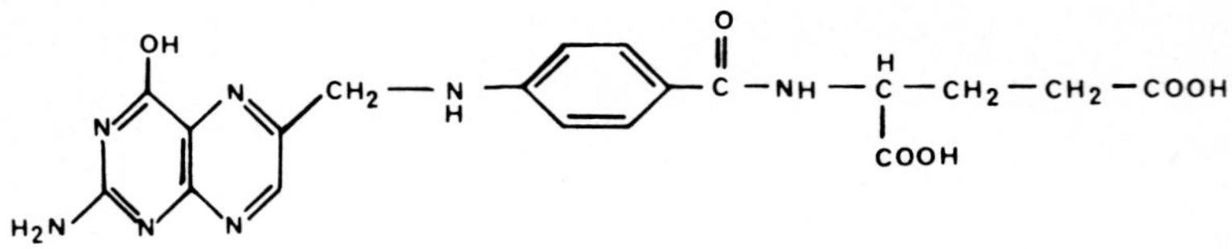

412. Coenzymes derived from the vitamin shown above are required by enzymes involved in the de novo synthesis of which of the following nucleotides?

(1) Adenosine triphosphate (ATP)
(2) Guanosine triphosphate (GTP)
(3) Thymidine triphosphate (TTP)
(4) Cytidine triphosphate (CTP)

413. Which of the following enzymes or proteins have porphyrin-containing coenzymes or prosthetic groups?

(1) Hemoglobin
(2) Methylmalonyl CoA mutase
(3) Cytochrome c
(4) Ferredoxin

414. Which of the following polypeptide hormones contain six amino acids in an S-S ring (with a disulfide bridge between the first and the last of the six amino acids)?

(1) Insulin
(2) Oxytocin
(3) Vasopressin
(4) Glucagon

SUMMARY OF DIRECTIONS

A	B	C	D	E
1, 2, 3 only	1, 3 only	2, 4 only	4 only	All are correct

415. Research to delineate the modes of action of hormones utilizes antimetabolites or specific blocking agents to inhibit reactions at certain steps. If puromycin and cycloheximide blocked the action of a certain hormone, and actinomycin D and mitomycin C did not, it could be assumed that the mode of action of the hormone required

(1) active membrane transport of specific amino acids
(2) DNA synthesis
(3) RNA synthesis
(4) protein synthesis

DIRECTIONS: The groups of questions in this section consist of five lettered headings followed by several numbered items. For each numbered item choose the **one** lettered heading with which it is **most** closely associated. Each lettered heading may be used once, more than once, or not at all.

Questions 416-418
For each hormone, select the endocrine gland which produces it.

    (A)   Adenohypophysis
    (B)   Adrenal cortex
    (C)   Adrenal medulla
    (D)   Neurohypophysis
    (E)   Thyroid

416. ACTH

417. Norepinephrine

418. Calcitonin

Questions 419-421
For each function, select the protein which governs it.

    (A)   Cholecystokinin
    (B)   Gastrin
    (C)   Insulin
    (D)   Intrinsic factor
    (E)   Secretin

419. Secretion of pancreatic juice into the intestine

420. Release of bile from the gallbladder

421. Secretion of acid in the stomach

Questions 422-424
For each fluid, choose its characteristic component.

    (A)   Calcium ion
    (B)   Bicarbonate ion
    (C)   Hydrogen ion
    (D)   Epinephrine
    (E)   Lysozyme

422. Pancreatic secretion

423. Gastric juice

424. Tears

Questions 425-427
For each of the steroid-derived compounds, choose the organ in which it is synthesized.

    (A)   Skin
    (B)   Bone
    (C)   Liver
    (D)   Kidney
    (E)   Small intestine

425. Cholecalciferol

426. 25-Hydroxycholecalciferol

427. 1,25-Dihydroxycholecalciferol

Membranes and Cell Structure

DIRECTIONS: Each question below contains five suggested answers. Choose the **one best** response to each question.

Questions 428-429

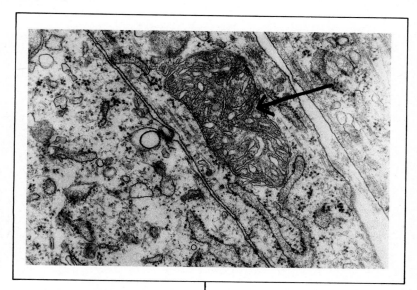

428. What is the structure shown in the electron micrograph above?

(A) Golgi apparatus
(B) Lysosome
(C) Mitochondrion
(D) Nucleus
(E) Rough endoplasmic reticulum

429. The structure indicated in the electron micrograph is involved primarily in

(A) energy production
(B) enzymatic degradation of ingested particles
(C) packaging of secretory proteins
(D) protein synthesis
(E) transmission of genetic information

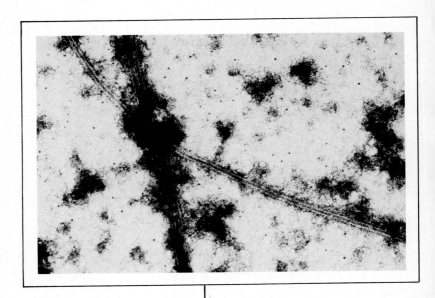

430. The electron micrograph above shows a protein structure found in mitotic spindles, eukaryotic flagella, and nerve axons. What is the structure?

(A) Ribosomes
(B) Actinomycin
(C) Microtubules
(D) Endoplasmic reticulum
(E) Microfilaments

431. A typical plasma membrane would be most likely to have which of the following weight compositions?

	Lipid	Protein	Carbohydrate	RNA
(A)	35%	45%	5%	10%
(B)	35%	55%	5%	0%
(C)	20%	75%	0%	0%
(D)	60%	30%	0%	5%
(E)	35%	40%	20%	0%

432. Which of the following compounds is an inhibitor of sodium-dependent glucose transport across the plasma membrane?

(A) Ouabain
(B) Sodium azide
(C) Dicumarol
(D) Phlorhizin
(E) Phloretin

433. The structure illustrated in the micrograph below is a

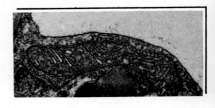

(A) ribosomal inclusion
(B) polyhedral body
(C) gas vacuole
(D) mitochondrion
(E) nucleolus

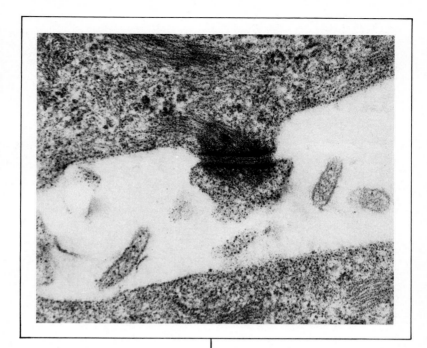

434. The membrane bridge illustrated in the electron micrography above is a

(A) septate junction
(B) gap junction
(C) desmosome
(D) zonula occludens
(E) conjugation junction

DIRECTIONS: Each question below contains four suggested answers of which **one** or **more** is correct. Choose the answer

A   if   **1, 2, and 3**   are correct
B   if   **1 and 3**   are correct
C   if   **2 and 4**   are correct
D   if   **4**   is correct
E   if   **1, 2, 3, and 4**   are correct

435. In which subcellular fractions does the chain elongation of fatty acids occur in the mammalian liver?

(1) Nuclear
(2) Mitochondrial
(3) Microsomal
(4) Supernatants of 100,000 x g (60 min)

436. The structure shown in the electron micrograph below is

(1) composed of protein and lipid bilayers
(2) void of enzymatic activity
(3) capable of transporting Na^+ against a chemical gradient
(4) the primary producer of intra-cellular energy

Photograph accompanies Question 436

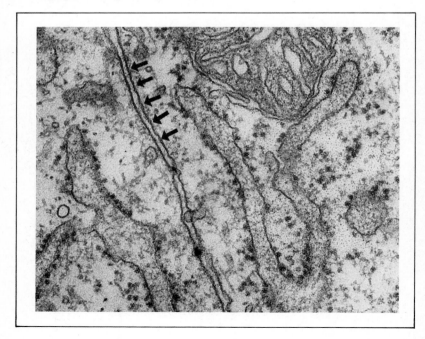

437. Crucial features of the unit-membrane model of Davson and Danielli include

(1) a bimolecular leaflet of lipids (bilayer)
(2) a hydrophobic interior and hydrophilic exterior
(3) a monolayer of protein at the aqueous interfaces of the membrane with its environment, i.e., the cytoplasm and the external milieu
(4) cholesterol molecules interspersed in the membrane to increase stability

438. Intracellular protein synthesis occurs

(1) on the rough endoplasmic reticulum
(2) in the Golgi apparatus
(3) on polysomes
(4) in the nucleus

439. Which of the following enzymes are found in lysosomes?

(1) Ribonuclease
(2) β-Galactosidase
(3) Acid phosphatase
(4) Cathepsins

440. Red blood cell ghosts shown in the electron micrograph below

(1) are devoid of cholesterol
(2) have lost their membrane bilayer
(3) contain internal mitochondrial vesicles (A)
(4) have lost mainly hemoglobin

Photograph accompanies Question 440

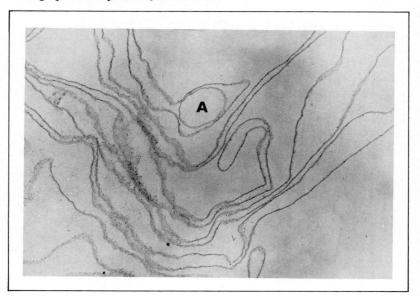

Questions 441-444

For each cellular structural element, choose its appropriate lettered location in the cell below.

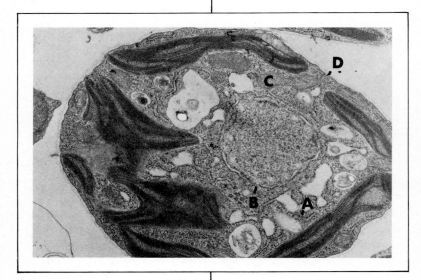

441. Endoplasmic reticulum

442. Nuclear pore

443. Mitochondria

444. Plasma membrane

For each function, select the lettered
area in the electron micrograph
below with which it is most likely to
be associated. Answer "E" if the
function is not association with any
of the designated areas.

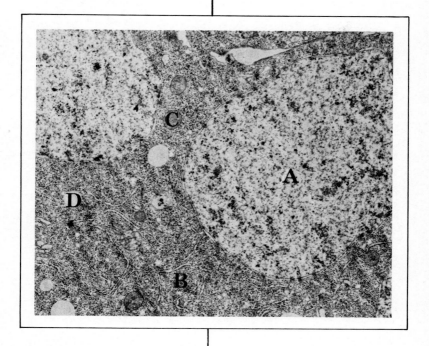

445. RNA synthesis

446. Oxidative phosphorylation

447. Protein synthesis

Metabolism

DIRECTIONS: Each question below contains five suggested answers. Choose the **one best** response to each question.

448. In facultative anaerobiosis, the Pasteur effect is the observation that the presence of oxygen

(A) increases the rate of gluconeo-genesis
(B) increases the rate of catabolism of glucose
(C) decreases the rate of catabolism of glucose
(D) decreases the amount of ATP formed per molecule of glucose from substrate phosphorylation
(E) decreases the rate of lipogenesis

449. The oxygen dissociation curve for hemoglobin is shifted to the right by

(A) decreased O_2 tension
(B) increased N_2 tension
(C) increased pH
(D) increased CO_2 tension
(E) decreased CO_2 tension

450. Which of the lettered points on the diagram below best represents completely compensated metabolic acidosis?

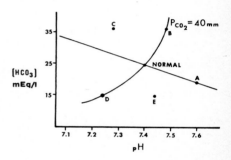

(A) A
(B) B
(C) C
(D) D
(E) E

451. A .22 M solution of lactic acid (pKa 3.9) was found to contain .20 M in the dissociated form, and .02 M undissociated. What is the pH of the solution?

(A) pH 2.9
(B) pH 3.3
(C) pH 3.9
(D) pH 4.9
(E) pH 5.4

452. The component in muscle tissue that contains the ATPase activity required for contraction is

(A) actin
(B) myosin
(C) sarcoplasmic reticulum
(D) motor end-plate
(E) calcium

453. The A antigen of the red blood cell is

(A) a fibrous protein
(B) a major surface lipoprotein
(C) reactive with the Kell blood group antigen
(D) a surface glycoprotein
(E) an endotoxin

454. Addition of which of the following compounds or enzymes to a cell-free cytoplasmic suspension would decrease levels of cyclic AMP?

(A) Cyclic AMP phosphodiesterase
(B) Dibutyryl cyclic AMP
(C) Caffeine
(D) Aminophylline
(E) Adenylate cyclase

455. The majority of energy for muscular contraction is stored in muscle tissue in the form of

(A) ADP
(B) phosphoenolpyruvate
(C) cyclic AMP
(D) ATP
(E) creatine phosphate

456. The pH of body fluids is stabilized by buffer systems. Which of the following compounds is the most effective buffer at physiologic pH?

(A) Citric acid, pKa = 3.09
(B) Na_2HPO_4, pKa = 12.32
(C) CH_3CO_2H, pKa = 4.74
(D) NH_4OH, pKa = 9.24
(E) NaH_2PO_4, pKa = 7.21

457. Water is 70 percent of body weight and may be said to be the "cell solvent." The most important property of water that accounts for its ability to dissolve compounds is

(A) the strong covalent bonds formed between it and salts
(B) the hydrogen bonds between it and biochemical molecules
(C) the absence of interacting forces among its molecules
(D) the hydrophobic bonds formed between it and long chain fatty acids
(E) its freezing point being well below body temperature

458. The normal osmotic pressure of proteins in the blood is approximately

(A) 0 mm Hg
(B) 10 mm Hg
(C) 30 mm Hg
(D) 70 mm Hg
(E) 90 mm Hg

459. Of the following, the prime, osmotic, regulatory component of plasma is

(A) immunoglobulin
(B) erythrocytes
(C) urea
(D) albumin
(E) glucose

460. Consider a suspension of mitochondria incubated with a respiratory substrate, e.g., succinate, plus phosphate. In this suspension, all of the following statements would be true of oxygen consumption EXCEPT

(A) it would be increased by the addition of ADP
(B) it would be increased by oligomycin if 2,4-dinitrophenol were present
(C) it would be increased by 2,4-dinitrophenol if oligomycin were present
(D) it would not be increased by ADP if oligomycin were present
(E) it would not be increased by ADP if 2,4-dinitrophenol were present

461. Which the schematic configurations below represents a stable lipid-water interaction?

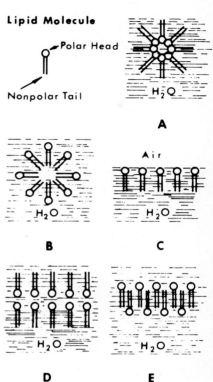

(A) Figure A
(B) Figure B
(C) Figure C
(D) Figure D
(E) Figure E

462. The plasma acid phosphatase level may be significantly elevated in

(A) metastatic prostatic carcinoma
(B) osteoblastic sarcoma
(C) obstructive jaundice
(D) Paget's disease
(E) rickets

463. Which of the following processes is characteristic only of the liver in mammals?

(A) Gluconeogenesis
(B) Hydroxylation of phenylalanine
(C) Glycogen synthesis and storage
(D) Epinephrine sensitivity
(E) Serum albumin synthesis

464. Which of the following blood proteins is the major source of extracellular cholesterol for the tissues in humans?

(A) Very low-density lipoprotein
(B) Low-density lipoprotein
(C) High-density lipoprotein
(D) Albumin
(E) γ-Globulin

465. Sandhoff's gangliosidosis is characterized by a deficiency of

(A) hexosaminidase A, only
(B) hexosaminidase B, only
(C) hexosaminidase C, only
(D) hexosaminidase A and B
(E) hexosaminidase A, B, and C

466. Which of the following is a sex-linked recessive disease?

(A) Lesch-Nyhan syndrome
(B) Infantile autism
(C) Maple syrup urine disease
(D) Familial hypercholesterolemia
(E) Cushing's syndrome

467. With a theoretical nonequilibrium situation, as shown below, and a membrane permeable only to sodium and chloride, what will be the final concentration of chloride on the left according to the Donnan equilibrium?

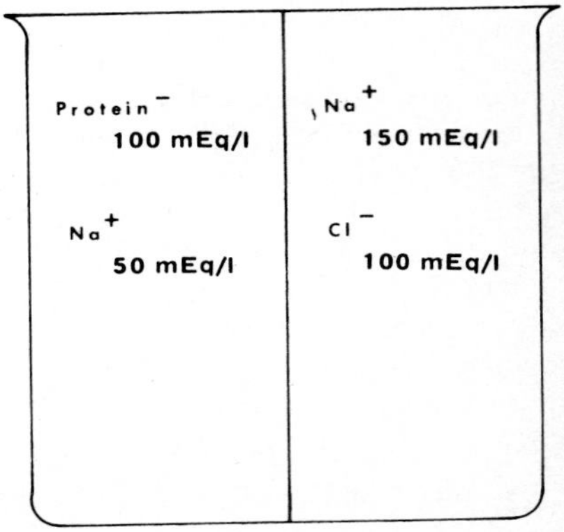

Protein$^-$	Na^+
100 mEq/l	150 mEq/l
Na^+	Cl^-
50 mEq/l	100 mEq/l

(A) 0
(B) 25 mEq/l
(C) 33 mEq/l
(D) 50 mEq/l
(E) 75 mEq/l

468. When fresh urine contains large amounts of porphobilinogen its color is

(A) normal
(B) pink to light red
(C) green
(D) dark red to brown
(E) black

80

469. The enzyme deficiency found in McArdle's syndrome (a glycogen storage disease) is

(A) pancreatic peptidase
(B) hepatic phosphorylase
(C) muscle phosphorylase
(D) hepatic glycogen synthetase
(E) debranching enzyme

470. The enzyme secreted by osteoblasts that is thought to provide phosphate for bone deposition is

(A) glucose 6-phosphate dehydrogenase
(B) acid phosphatase
(C) adenosine triphosphatase
(D) alkaline phosphatase
(E) phosphorylase kinase

471. Phenylketonuria is caused by a lack of

(A) phenylalanine hydroxylase
(B) phenylalanine α-ketoglutaric transaminase
(C) homogentisate oxidase
(D) Dopa decarboxylase
(E) cystathionine synthetase

472. Which of the following enzymes is deficient in von Gierke's disease?

(A) Uridine diphosphate glucose pyrophosphorylase
(B) Glucose 6-phosphatase
(C) α-1,6-Glucosidase
(D) Glucokinase
(E) Phosphoglucomutase

473. Which of the following diseases is NOT caused by a functionally defective enzyme?

(A) Phenylketonuria
(B) Tay-Sachs disease
(C) Von Gierke's disease
(D) Gaucher's disease
(E) Caisson disease

474. In type IV hereditary disorders of glycogen metabolism (Andersen's disease) an abnormal glycogen with few branches is formed and stored in the liver. The absence of which enzyme is responsible for this condition?

(A) Glucose 6-phosphate
(B) Amylo-1,6-glucosidase
(C) Amylo-(1,4 → 1,6)-transglycosylase
(D) Glycogen phosphorylase
(E) Phosphorylase kinase

475. Of the following body fluids, the one with the lowest pH is

(A) plasma
(B) pancreatic juice
(C) liver bile
(D) gastric juice
(E) sweat

DIRECTIONS: Each question below contains four suggested answers of which **one** or **more** is correct. Choose the answer

A	if	**1, 2, and 3**	are	correct
B	if	**1 and 3**	are	correct
C	if	**2 and 4**	are	correct
D	if	**4**	is	correct
E	if	**1, 2, 3, and 4**	are	correct

476. Which of the following statements about the sequence and control of metabolic functions are true?

(1) For a given molecule its anabolic path is usually the reverse of the catabolic process
(2) Anabolism allows the synthesis of macromolecules from small precursors and requires energy input
(3) A given set of enzymes is usually used in only one metabolic pathway to allow for independent control mechanisms
(4) The initial reactions of most metabolic sequences are thermodynamically irreversible and are often sites for metabolic control

477. Which of the following substances decrease the affinity of hemoglobin for oxygen?

(1) 2,3-Diphosphoglycerate
(2) Carbon dioxide
(3) Hydrogen ions
(4) Hydroxide ions

478. During starvation, an increase in the activity of which of the following hepatic enzymes is expected?

(1) Hexose monophosphate shunt enzymes
(2) Lipogenesis enzymes
(3) Glycolysis enzymes
(4) Gluconeogenesis enzymes

479. All of the following compounds are synthesized in mammalian tissue EXCEPT

(1) biotin
(2) choline
(3) dehydroshikimic acid
(4) inositol

480. Damage to which of the following tissue cells elevates serum alkaline phosphatase levels?

(1) Biliary tract
(2) Osteoblasts
(3) Small intestine mucosa
(4) Placenta

481. Which of the following can NOT oxidize glucose?

(1) Liver
(2) Brain
(3) Heart
(4) Erythrocytes

482. Which of the following are capable of gluconeogenesis?

(1) Erythrocytes
(2) Kidney
(3) Skeletal muscle
(4) Liver

483. A patient with tyrosinosis

(1) excretes para-hydroxyphenyl-pyruvic acid
(2) is unable to utilize homo-gentisic acid
(3) lacks an oxidase
(4) has elevated levels of phenyl-lactic acid

484. Phenylketonuria is thought to be a genetic disease related to

(1) a constitutive mutation in the synthesis of shikimic acid
(2) tyrosinase deficiency
(3) deficiency of a phenylalanine transport enzyme
(4) deficiency of a hydroxylase in an amino acid's metabolism

485. In maple syrup urine disease, the α-ketoacids that accumulate are derived from which of the following amino acids?

(1) Leucine
(2) Lysine
(3) Isoleucine
(4) Phenylalanine

486. Urine that darkens on stand-ing may contain

(1) metanephrine
(2) porphobilinogen
(3) 5-hydroxyindoleacetic acid
(4) homogentisic acid

487. The inability to utilize glycogen to produce energy is associated with which of the fol-lowing enzyme deficiencies?

(1) Amylo-1,6-glucosidase
(2) Phosphorylase
(3) Glucose 6-phosphatase
(4) Amylo-(1,4 → 1,6)-transglyco-sylase

488. An abnormal glucose tolerance test may occur in patients with

(1) dumping syndrome
(2) rheumatoid arthritis
(3) oat-cell carcinoma
(4) diabetes

489. Once a diabetic patient is started on chronic insulin therapy

(1) serum insulin levels can no longer be measured accurately
(2) impotence occurs within a few months
(3) insulin requirements may decrease to nothing for a period of time before increasing
(4) a diet without any restrictions may be completely tolerable

490. Heavy alcohol ingestion may be associated with

(1) fatty liver change
(2) polyneuropathy
(3) ketoacidosis
(4) Wernicke-Korsakoff syndrome

491. Which of the following are capable of oxidizing ketone bodies to carbon dioxide?

(1) Erythrocytes
(2) Brain
(3) Liver
(4) Heart

492. Which of the following are capable of synthesizing ketone bodies from fatty acids?

(1) Erythrocytes
(2) Brain
(3) Skeletal muscle
(4) Liver

DIRECTIONS: The groups of questions in this section consist of four or five lettered headings followed by several numbered items. For each numbered item choose the **one** lettered heading with which it is **most** closely associated. Each lettered heading may be used once, more than once, or not at all.

Questions 493-496
For each enzyme, choose the disease its deficiency is most likely to produce.

(A) Hartnup's disease
(B) Albinism
(C) Crigler-Najjar syndrome
(D) Hemolytic anemia
(E) Wilson's disease

493. Tryptophan pyrrolase

494. Uridine diphosphate glucuronate transferase

495. Ceruloplasmin

496. Pyruvate kinase

Questions 497-500
Match the following. Each lettered heading may be used once, more than once, or not at all.

(A) Juvenile diabetes
(B) Adult onset diabetes
(C) Both
(D) Neither

497. Usually an absolute need for insulin

498. Frequent ketoacidosis

499. Unstable and "brittle" diabetes

500. Insulin resistance

Answers, Explanations
and References

Amino Acids, Proteins, and Enzymes

1. The answer is A. *(White, ed 5. p 92.)* The molar extinction coefficients of tryptophan, tyrosine, and phenylalanine at 280 nm are 5500, 540, 120, respectively. The absorbance maximum of peptide bonds is 215 nm.

2. The answer is B. *(White, ed 5. pp 96-97.)* Because glycine has two hydrogen atoms bound to its α-carbon, it is without right- and left-handed forms and, therefore, optically inactive. It is the only optically inactive amino acid naturally found in proteins.

3. The answer is D. *(Mahler, ed 2. pp 43-49, 91-95.)* At pH 11, glu-his-arg-val-lys-asp has a net negative charge. Therefore, in an electric field, the peptide will move toward the anode at pH 11.

4. The answer is C. *(Mahler, ed 2. p 14.)* The figure shows the titration curve of glycine. The maximum buffering capacity of any ionizable function is at the pH equivalent to the pK represented by points A and B.

5. The answer is A. *(White, ed 5. pp 119-120.)* Proteins that contain many ionizable groups are effective intracellular buffers. The imidazolium group of histidine has a pK of approximately 6.5. Histidine is the only amino acid with a side chain pK approximately at neutrality.

6. The answer is C. *(White, ed 5. pp 90-95, 97-106.)* The epsilon-amino group of lysine carries a positive charge. The carboxylate side chain of glutamic acid is negative, and the side chains of the remaining amino acids mentioned are neutral.

7. The answer is B. *(Mahler, ed 2. p 11.)* The acidity of aqueous solutions is expressed as the pH and is equal to the negative logarithm to the base 10 of the hydrogen ion concentration.

8. The answer is E. *(White, ed 5. p 107.)* Ninhydrin reacts with amino acids to yield decarboxylation and deamination of alpha-amino acids while it is reduced to hydrindantin. One molecule each of the resulting ammonia, ninhydrin, and hydrindantin complex to form a purple pigment with an optical absorbance maximum of about 570 nm.

9. The answer is **E.** *(White, ed 5. p 107.)* The Millon reaction [Hg $(NO_3)_2$ in nitric acid with a trace of nitrous acid] is a test for phenolic compounds. Tyrosine reacts in this test to yield a red color. The Pauly reaction with diazotized sulfanilic acid in alkaline solution would detect tyrosine and histidine as well.

10. The answer is **C.** *(Mahler, ed 2. pp 775-777.)* Hydroxyproline is not directly incorporated into polypeptides. Prolyl residues already incorporated in peptide linkage are hydroxylated by a reaction utilizing molecular oxygen to give hydroxyprolyl residues which are found exclusively in collagen.

11. The answer is **A.** *(White, ed 5. pp 635-636.)* Alanine is the transamination derivative of pyruvic acid. Serine is formed from glycolytic intermediates. The remaining amino acids on the list are dietary essentials which are not synthesized by man.

12. The answer is **A.** *(White, ed 5. pp 550-551, 688-689.)* The final oxidation products of isoleucine are acetic and propionic acids. These acids, if not further oxidized to CO_2, are the precursors of nonessential, even- and odd-chain fatty acids, respectively.

13. The answer is **C.** *(White, ed 5. pp 96, 1032.)* Thyroxine is a derivative of tyrosine, obtained by iodination and coupling of tyrosyl moieties in the thyroid.

14. The answer is **C.** *(White, ed 5. p 640.)* Because the enzyme phenylalanine hydroxylase exists in the normal mammalian liver and kidney, phenylalanine "spares" the dietary requirement for tyrosine. The other amino acids on the list must be supplied by diet.

15. The answer is **D.** *(White, ed 5. p 720.)* Aspartate and carbamoyl phosphate are condensed to give carbamoyl aspartate in one of the early steps of pyrimidine biosynthesis.

16. The answer is **E.** *(Mahler, ed 2. p 47.)* The structure shown is the amino acid tryptophan.

17. The answer is **C.** *(White, ed 5. pp 172-174.)* Amino acids, such as tryptophan, are the components of proteins, e.g., hemoglobin.

18. The answer is **C.** *(Mahler, ed 2. p 799.)* Leucine is the only amino acid which is ketogenic but not glycogenic. Isoleucine, lysine, phenylalanine, and tyrosine are both glycogenic and ketogenic. All other amino acids are glycogenic.

19. The answer is C. *(White, ed 5. pp 677-678.)* While other amino acids which are ketogenic may also be employed for gluconeogenesis, leucine is exclusively ketogenic.

20. The answer is B. *(White, ed 5. pp 801-803.)* The plasma concentrations of the following amino acids are given in mg/100 ml: glutamine 4.5 − 10.0; alanine 2.5 − 7.5; glycine 0.8 − 5.4; and histidine 0.8 − 3.8.

21. The answer is B. *(White, ed 5. pp 690-691.)* Phenylalanine and tyrosine are catabolized via homogentisic acid to 4-maleylacetoacetic acid. This product is cleaved to fumarate and acetoacetate.

22. The answer is B. *(Mahler, ed 2. p 807.)* Homocysteine is methylated by N^5-methyltetrahydrofolate to generate methionine. This methyl group may then be transferred to other compounds by S-adenosylmethionine.

23. The answer is C. *(Mahler, ed 2. p 807.)* Norepinephrine is transmethylated to form epinephrine by S-adenosylmethionine which is left S-adenosylhomocysteine.

24. The answer is C. *(White, ed 5. p 691.)* Decarboxylation of serine yields carbon dioxide and ethanolamine. Ethanolamine has a major role as a constituent of the cephalins, and via choline of the lecithins (phosphatides).

25. The answer is A. *(White, ed 5. pp 669-670.)* Decarboxylation of aspartate at the gamma position produces alanine. Alanine is also produced by transamination of pyruvate.

26. The answer is A. *(White, ed 5. pp 639, 644-645.)* Homocysteine is a sulfur-containing amino acid that is not found in proteins. It lies between S-adenosylhomocysteine and homocysteine, cystathionine, and methionine in sulfur metabolism.

27. The answer is C. *(White, ed 5. p 670.)* Decarboxylation of histidine yields histamine, a compound with vasodilating activity and a role in mediating allergy.

28. The answer is C. *(White, ed 5. p 650.)* The enzyme which condenses aspartate with citrulline to yield argininosuccinate, hydrolyzes ATP to adenosine monophosphate (AMP) and pyrophosphate.

29. The answer is E. *(White, ed 5. pp 649-650.)* In the urea cycle, ornithine is carbamoylated to yield citrulline which condenses with aspartate to yield argininosuccinate. Argininosuccinase catalyzes the formation of arginine which in turn is hydrolyzed to form ornithine and urea.

30. The answer is A. *(Mahler, ed 2. p 778.)* In *E. coli* there are three aspartokinase isozymes, each sensitive to inhibition by a different amino acid (threonine, methionine, and lysine). Each isozyme is thought to play a role in the biosynthesis of the amino acid which inhibits that isozyme.

31. The answer is C. *(Mahler, ed 2. pp 133-145.)* The α-helix is one type of periodic or secondary structure found in some proteins. Among the stabilizing forces are the intramolecular hydrogen bonds formed between peptide bonds and the geometric arrangement of the helix that leads to a minimum of unfavorable steric interactions among the R groups of the amino acid residues. Hydrophobic interactions are not thought to be important in stabilizing α-helical structure, and regions of the α-helix do tend to be interrupted by prolyl or glycyl residues.

32. The answer is A. *(Mahler, ed 2. p 174.)* Isozymes are enzymes that perform the same function, but occur in more than one molecular form within the same tissue or species. This phenomenon may result from variable assembly of a number of different monomeric subunits into distinct multimeric complexes.

33. The answer is E. *(Mahler, ed 2. pp 139-140.)* Figure A is a random coil; Figure B an α-helix; Figure C a pleated sheet; and Figure D a triple-stranded helix. Figure E is a super coiled helix, one of the forms of hair keratin.

34. The answer is E. *(Mahler, ed 2. pp 139-140.)* Tropocollagen has the structure of a triple helix because many proline and hydroxyproline residues prevent the hydrogen bond formation necessary for an α-helix.

35. The answer is D. *(White, ed 5. p 438.)* Assuming LDH is a tetrameric protein formed by random association of two electrophoretically distinct polypeptides, A and B, one can predict the existence of five isozymes: A_4, A_3B_1, A_2B_2, A_1B_3, B_4.

36. The answer is E. *(Mahler, ed 2. pp 762-769.)* Pepsinogen, procarboxypeptidase, chymotrypsinogen, and trypsinogen are inactive zymogen precursors of the gastrointestinal enzymes pepsin, carboxypeptidase, chymotrypsin, and trypsin respectively. Ribonuclease is secreted in its active form.

37. The answer is E. *(White, ed 5. pp 146-147, 251, 630-632.)* Trypsin is an endopeptidase which hydrolyzes the carboxyl peptide linkages of arginine and lysine which are both positively charged amino acids. Chymotrypsin is specific for the carboxyl peptide linkages of aromatic amino acids. Carboxypeptidase hydrolyzes amino acids sequentially from the carboxyl end of proteins. Renin produces paracasein from casein, allowing its degradation by pepsin into proteoses and peptones.

38. **The answer is D.** *(White, ed 5. pp 949-951.)* Two molecules of actin are bound to myosin on each of the two globular heads of the molecule. These heads contain the adenosine triphosphatase (ATPase) activity.

39. **The answer is B.** *(White, ed 5. pp 808-814.)* Immunoglobulins contain light and heavy chains linked by disulfide bonds and a small amount of carbohydrate. Immunoglobulins are synthesized by lymphocytes and plasma cells in response to specific antigens, and can be differentiated into at least five classes, i.e., G, A, M, D, and E.

40. **The answer is B.** *(White, ed 5. p 859.)* The catabolism of hemoglobin first involves scission of the α-methene bridge to yield choleglobin. Removal of the apoprotein yields verdohemochrome. Verdohemochrome is converted to biliverdin which, in turn, is converted to bilirubin.

41. **The answer is A.** *(White, ed 5. pp 860-861.)* Stercobilin is the pigment which imparts the characteristic color to stools. Bilirubin diglucuronide is hydrolyzed to yield bilirubin which is reduced to urobilinogen by bacterial flora. Urobilinogen is further reduced to stercobilinogen. Oxidation of stercobilinogen in air yields stercobilin.

42. **The answer is C.** *(Mahler, ed 2. p 277.)* The reciprocal of velocity is plotted on the Y axis against the reciprocal of substrate concentration on the X axis.

43. **The answer is B.** *(Mahler, ed 2. pp 295-297.)* Increasing the proton concentration would decrease the V_{max} without affecting K_m. Therefore, the inhibition pattern would be noncompetitive.

44. **The answer is A.** *(Mahler, ed 2. pp 295-299.)* Competitive inhibition kinetics are defined as effects that increase K_m of an enzyme without affecting V_{max}.

45. **The answer is B.** *(White, ed 5. p 233.)* Catalysts, of which enzymes are examples, lower the activation energy of a reaction without affecting the energy of reactants or products.

46. **The answer is C.** *(White, ed 5. pp 242-244.)* The intercept of the double-reciprocal plot on the abscissa is $-1/K_m$. Since $-1/K_m = -3$, $K_m = 0.33$.

47. **The answer is A.** *(White, ed 5. pp 210-214, 280.)* Assuming that

$$K_{eq} = \frac{(C)}{(A)(B)} = 0.1,$$

then $\Delta F^\circ = -RT\,(2,3) \log K_{eq} = -2.3\,RT \log(0.1) = 2.3\,RT.$

91

48. **The answer is B.** *(White, ed 5. pp 236-238.)* As shown on a double-reciprocal plot, a competitive inhibitor which increases K_m without affecting V_{max}, would increase the slope without affecting the intercept on the ordinate.

49. **The answer is E.** *(White, ed 5. pp 224-230.)* The ordinate intercept of the double-reciprocal plot of Michaelis-Menten enzyme reaction kinetics gives the value of $1/K_m$.

50. **The answer is E.** *(White, ed 5. pp 212, 280-281.)* Because free energy is a thermodynamic function of state, overall energy changes in biochemical reactions are independent of the reaction path.

51. **The answer is C.** *(White, ed 5. p 317.)* The redox system with the largest standard reduction potential among those listed is cytochrome a (Fe^{3+}/Fe^{2+}). Electrons are passed from carriers with low standard reduction potentials to those with high standard reduction potentials.

52. **The answer is D.** *(White, ed 5. p 211.)* The law is concerned with the concept of entropy, not free energy. A closed system proceeds spontaneously to increased entropy and disorder.

53. **The answer is D.** *(White, ed 5. pp 364-365, 367-370.)* Uncoupling agents allow for continued mitochondrial respiration without concomitant phosphorylation of ADP. Uncouplers stimulate the mitochondrial ATPase.

54. **The answer is C.** *(White, ed 5. pp 367-368.)* Dinitrophenol is an uncoupler of oxidative phosphorylation.

55. **The answer is E.** *(White, ed 5. pp 242-247.)* For homotropic enzyme molecules, each substrate molecule bound causes a conformational change in the enzyme, enhancing subsequent substrate binding and reaction velocity.

56. **The answer is D.** *(Davis, ed 2. p 127.)* Polymyxin, with one polar head and an aliphatic tail, resembles a cationic detergent and disrupts bacterial cell walls. It causes cell walls to become leaky, to lose intracellular constituents, and to admit normally excluded dyes.

57. **The answer is A.** *(Davis, ed 2. pp 112, 120.)* D-cycloserine is a structural analog of D-alanine. It blocks dipeptide synthetase to make D-alanyl-D-alanine. This competitive inhibition may be reversed by increasing the concentration of D-alanine.

58. The answer is B. *(Anderson, ed 6. p 176.)* Fluorescent techniques have demonstrated localization of tetracyclines within mitochondria. The tetracyclines are chelators of divalent cations, and are useful in measuring the activity of bone, e.g., in osteomalacia.

59. The answer is B. *(White, ed 5. pp 774-780.)* A constitutive enzyme is one which is synthesized under all physiologic conditions. Its enzyme levels are dependent on neither an inducer nor a co-repressor.

60. The answer is D. *(White, ed 5. pp 775-776.)* The term induction is applied in the case where an enzyme's substrate leads to the de novo synthesis of the enzyme. Induction can be demonstrated with β-galactosidase in *E. coli,* where the inducer and substrate function can be separated and have been analyzed.

61. The answer is D. *(Lehninger, 1970. p 739.)* Most vertebrate organs are provided with a constant environment and do not need to respond to sudden changes. The liver, however, is exposed to varying rates of incoming nutrients and must respond rapidly to changes by induction or derepression of its enzymes.

62. The answer is C. *(Mahler, ed 2. p 858.)* Only a mutational block in enzyme C would produce an absence of Z while the pathway continued through M to L.

63. The answer is A (1, 2, 3). *(Mahler, ed 2. pp 57-59.)* Because of its partial double-bond character, there is not free rotation about the peptide bond. The peptide bond usually has a *trans* configuration and is shorter than a carbon-carbon single bond.

64. The answer is E (all). *(Mahler, ed 2. pp 43-49.)* Isoleucine, methionine, proline, and phenylalanine are all amino acids with relatively nonpolar, hydrophobic side chains.

65. The answer is A (1, 2, 3). *(Mahler, ed 2. pp 772-773.)* Essential amino acids are those which must be supplied in the diet for normal growth, development, and health. While the dietary requirements for amino acids may be altered by various physiologic states, it is generally agreed that the essential amino acids are leucine, isoleucine, valine, lysine, methionine, phenylalanine, tryptophan, threonine, and arginine. While histidine is essential in the rat, it is alleged that humans are not dependent on dietary histidine.

66. The answer is A (1, 2, 3). *(McGilvery, ed 1. pp 422-424, 462-471, 474-478, 495-500.)* Glycine is incorporated into heme, creatine, and guanine. There is no pathway for the direct incorporation of glycine into thymine.

67.   The answer is C (2, 4). *(McGilvery, ed 1. pp 412-414,424, 429, 662-663.)* Methionine and tryptophan are essential amino acids and, therefore, cannot be synthesized by mammalian tissues. Glycine and choline can, however, be synthesized by mammalian tissues with serine as a precursor.

68.   The answer is C (2, 4). *(Davis, ed 2. p 156.)* 6-Aminopenicillanic acid may be considered to be the condensation product of L-cysteine and D-valine. The 6-amino group receives a large variety of acyl groups, giving rise to a spectrum of penicillins.

69.   The answer is D (4). *(Davis, ed 2. pp 120, 156.)* Cycloserine is essentially a closed alanine ring. In bacterial extracts cycloserine inhibits the enzyme that racemizes L-alanine to yield D-alanine, and also the enzyme which incorporates D-alanine into dipeptides: it thus blocks two early steps in cell wall synthesis.

70.   The answer is B (1, 3). *(Davis, ed 2. pp 154-155, 159.)* Sulfisoxazole, a sulfonamide analog, and p-aminosalicyclic acid inhibit folate metabolism. Isoniazid blocks NAD-requiring steps and enzymes requiring pyridoxal phosphate cofactors. Tetracycline causes a bacteriostatic inhibition of protein synthesis.

71.   The answer is D (4). *(Mahler, ed 2. pp 164-170.)* In general, it is the primary structure that determines the higher order structure of proteins. (Of course, there are certain well known exceptions, such as insulin.) Disulfide bonds are usually of more importance in stabilizing than in determining a protein's conformation. While charged amino acids tend to be exposed to water, hydrophobic side chains tend to be buried within the center of the molecule away from water.

72.   The answer is A (1, 2, 3). *(Mahler, ed 2. pp 129-133.)* Gamma immunoglobulins are composed of two heavy and two light chains with intra- and inter-chain disulfide linkages. There are two antigen binding sites per molecule. The N-terminus of both heavy and light chains is the variable region, whereas the C-terminus is constant. In multiple myeloma, large quantities of a homogeneous light chain (Bence Jones protein) may be excreted in the urine.

73.   The answer is D (4). *(Mahler, ed 2. pp 762-769.)* Trypsinogen and chymotrypsinogen are synthesized in the pancreas. They are the zymogen precursors of the endopeptidases trypsin and chymotrypsin, respectively. Limited digestion with trypsin converts the zymogens into the active enzymes. The two zymogens have considerable homology in amino acid sequence.

74. The answer is E (all). *(Mahler, ed 2. pp 325-376.)* Most enzymes are proteins with a molecular weight in excess of 5000 daltons. They have a high degree of specificity for substrate and catalyze a specific reaction. They hasten the reaction by decreasing its activation energy. Most, but not all, enzymes have optimal activity in the vicinity of neutral or physiologic pH.

75. The answer is B (1, 3). *(White, ed 5. pp 144, 146, 261-263.)* Fragment A, consisting of the light chain and part of the heavy chain, contains the antibody combining site and the "hypervariable regions." Fragment A is known as $F_{(ab)}^2$ when B is known as $F_{(c)}$.

76. The answer is E (all). *(Mahler, ed 2. pp 419-420.)* The various porphyrins are derived from a tetrapyrrole ring nucleus. They are able to form metal chelates with many ions including iron, copper, zinc, and cobalt. Protoporphyrin IX complexed with iron is referred to as heme and is found in hemoglobin, myoglobin, cytochrome, and catalase.

77. The answer is E (all). *(White, ed 5. pp 159, 171, 178, 179, 392-394.)* Hemoglobin, myoglobin, and cytochrome c all contain iron in a heme ring. Peroxidase also contains iron, but without the heme moiety.

78. The answer is A (1, 2, 3). *(White, ed 5. pp 830-832.)* The complement system is a cascade usually activated by antigen-antibody complexes. It lyses red cells, and produces chemotactic peptides.

79. The answer is E (all). *(White, ed 5. pp 242-243.)* According to the Monod model of allosteric enzymes, such enzymes consist of polymers of subunits each containing both an active catalytic site and an allosteric site for the binding of modifiers. It further assumes that conformational changes occur in a concerted fashion, i.e., all subunits in a given polymer are in the same state at the same time. In a K-system, the different conformational states of the subunits differ with respect to their affinities for various ligands, i.e., substrates, inhibitors, or activators.

80. The answer is B (1, 3). *(Orten, ed 8. p 129.)* When a modifier binds at the allosteric site it affects the active site by altering V_{max} and K_m. Of course, binding of both substrate and modifier are concentration-dependent. The velocity of the reaction depends on the concentrations of both the substrate and the modifier.

81. The answer is C (2, 4). *(Orten, ed 8. p 129.)* Examples of allostery may include enzymes, e.g., *E. coli* aspartate transcarbamoylase, or binding proteins, e.g., hemoglobin. Nonregulatory enzymes, such as phosphoglucose isomerase and lactate dehydrogenase, do not exhibit sigmoidal kinetics.

82. The answer is D (4). *(Mahler, ed 2. p 33.)* Entropy is a measure of the degree of order in a system. Enthalpy, free energy, and equilibrium constants are indirect parameters of order in a biologic system.

83. The answer is B (1, 3). *(Mahler, ed 2. pp 666-671.)* Hemoglobin is a tetrameric hemoprotein which binds oxygen with sigmoidal kinetics because of cooperative interactions among the four binding sites. Oxygen is bound to hemoglobin without changing the redox state of the iron from the ferrous state. Carbon monoxide and cyanide both bind to hemoglobin more tightly than does oxygen itself.

84. The answer is B (1, 3). *(White, ed 5. pp 360-370.)* Two of the hypotheses which attempt to explain the mechanism of oxidative phosphorylation are the chemiosmotic hypothesis of Mitchell and the chemical coupling hypothesis. The Pasteur effect is the inhibition of glycolysis by oxygen which Warburg's hypothesis related to the metabolism of tumors.

85. The answer is B (1, 3). *(Mahler, ed 2. pp 27-32.)* The oxidation-reduction potential is a quantitative measure of the tendency of a compound to donate or accept electrons. The hydrogen electrode is arbitrarily assigned a standard potential of zero. Compounds with negative reduction potentials will tend to reduce protons to hydrogen gas, while compounds with positive reduction potentials will tend to accept electrons from H_2. Standard redox potentials apply to the situations where all reactants and products are present at concentrations of 1 M or tensions of 1 atm; it may be necessary to correct for actual concentrations. The standard free energy change of a reaction can be calculated from the standard reduction potential by the formula $\Delta F^\circ = -nFE^\circ$.

86. The answer is B (1, 3). *(Mahler, ed 2. pp 680-684.)* The standard redox potential of cytochrome b is lower than that of either cytochrome c or a. Cytochrome does not interact directly with oxygen, cyanide, or carbon monoxide. It reacts with cytochrome a through cytochrome c.

87. The answer is C (2, 4). *(Mahler, ed 2. pp 690-692.)* There are thought to be three sites of energy conservation in the mitochondrial electron transport chain that lead to the production of ATP: between NAD and a flavoprotein (site I); between cytochrome b and cytochrome c (site II); and between cytochrome c and cytochromes a, a_3 (site III).

88. The answer is B (1, 3). *(White, ed 5. pp 868-869.)* Sickle cell anemia results from a single amino acid substitution in the hemoglobin beta-chain, and occurs in varying degrees in persons with S-S homozygous genetic constitution. Individuals with a single allele for hemoglobin S (heterozygotes) have sickle cell trait; and though they may have a mild microcytic anemia, they do not have sickle cell disease. The homozygous state confers protection against malarial parasites.

89. **The answer is E (all)** *(Davis, ed 2. pp 135-136, 163, 357.)* The two bacterial enzymes and the transport protein listed are inducible. Immunoglobulin synthesis is inducible by exposure to antigen to which the cell has been sensitized.

90. **The answer is B (1, 3).** *(Davis, ed 2. pp 313-319.)* The feedback inhibition of allosteric enzymes by synthetic products is competitively antagonized by substrate or substrate analogs, even when the two compounds differ markedly in structure. The substrate and product are bound to different sites on the enzyme.

91. **The answer is C (2, 4).** *(White, ed 5. pp 821, 827, 830.)* Heparin is a naturally occurring mucopolysaccharide that prevents blood clotting by interference with a number of steps in the coagulation cascade. Plasmin is a circulating enzyme that hydrolyzes fibrin clots to form soluble products. Fibrinogen is the substrate that is acted upon by thrombin to yield the fibrin mesh of blood clots. Platelets are the agents that are first activated at the site of disruption in blood vessels to prevent hemorrhage.

92. **The answer is E (all).** *(Davis, ed 2. pp 131-132.)* All of the characteristics listed apply to the galactoside transport system of *E. coli.*

93. **The answer is D (4).** *(Davis, ed 2. pp 409-411.)* Heavy and light chains of IgG may be separated by reduction of interchain disulfide bonds with mercaptoethanol. Papain digests IgG into three proteolytic fractions (Fc and two Fab portions), while pepsin treatment yields almost indistinguishable $(Fab')_2$ fragments and digests completely the Fc fragment.

94-95. **The answers are: 94-C, 95-A.** *(Mahler, ed 2. pp 775-789.)* Histidine is synthesized (in microorganisms) from ribose 5-phosphate as an early precursor. Proline is synthesized in animal tissues, as well as in microorganisms, from glutamate as a precursor.

96-99. **The answers are: 96-D, 97-B, 98-E, 99-C.** (Mahler, ed 2. pp *43-49.)* Cysteine contains sulfur. Lysine is a basic amino acid. The side chain of leucine is an aliphatic, branched chain. Phenylalanine has an aromatic side chain.

100-102. **The answers are: 100-C, 101-D, 102-B.** *(White, ed 5. pp 117, 670-671, 978-979.)* Hair and nails are composed of keratin which contains a large amount of the disulfide amino acid cystine (approximately 14 percent in human hair). The enzyme tyrosinase catalyzes a series of reactions leading from tyrosine to dihydroxyphenylalanine and to melanin, which is a skin and hair pigment. Twenty percent of the amino acids in collagen, a triple-stranded protein formed by fibroblasts, is proline and hydroxyproline.

103-104. **The answers are: 103-D, 104-A.** *(Mahler, ed 2. pp 115-118.)* Mercaptoethanol reduces disulfide linkages to sulfhydryl groups. Chymotrypsin hydrolyzes peptide linkages where the carboxyl function is donated by an aromatic amino acid.

105-107. **The answers are: 105-D, 106-C, 107-E.** *(Mahler, ed 2. p 117.)* Trypsin is an endopeptidase which preferentially cleaves after basic amino acids (lysine or arginine). Chymotrypsin is an endopeptidase which preferentially cleaves after aromatic amino acids. Carboxypeptidase is an exopeptidase which cleaves at the carboxyl terminus of a polypeptide.

108-110. **The answers are: 108-A, 109-C, 110-D.** *(Mahler, ed 2. pp 267-282.)* The V_{max} of the enzyme is $k_3 E_o$. The dissociation constant of the enzyme-substrate complex and the Michaelis constant are k_2/k_1 and $(k_2 + k_3)/k_1$, respectively.

Nucleic Acids and Molecular Genetics

---

111. The answer is D. *(Harper, ed 15. p 50.)* Adenine, guanine, and cytosine would be incorporated into both DNA and RNA. Uracil is found only in RNA and thymine only in DNA.

112. The answer is D. *(Harper, ed 15. p 50.)* While thymine is the base which is paired with adenine in DNA, uracil is found in mRNA as the complement of thymine in the template DNA.

113. The answer is C. *(Harper, ed 15. p 51.)* Up to 10 percent of the nucleotides of tRNA have unusual methylated bases. Mitochondrial and nucleolar DNAs contain conventional bases. Nucleolar DNA codes for ribosomal rRNA.

114. The answer is A. *(White, ed 5. p 718.)* Lesch-Nyhan syndrome is caused by the absence of the enzyme hypoxanthine-guanine phosphoribosyltransferase (HGPRT). Among the symptoms of Lesch-Nyhan syndrome are hyperuricemia and gout. Other features of the syndrome such as self-mutilation are less well explained by HGPRT deficiency.

115. The answer is E. *(Mahler, ed 2. pp 806-807, 833-834.)* S-Adenosylmethionine donates methyl groups in the biosynthesis of creatine phosphate, epinephrine, melatonin, and phosphatidylcholine. Thymine, however, receives a methyl group from N^5, N^{10}-methylenetetrahydrofolate.

116. The answer is B. *(White, ed 5. pp 644-645.)* S-Adenosylmethionine donates the methyl group of methionine in many biochemical reactions leaving S-adenosylhomocysteine. These reactions include methylation of guanidoactic acid to yield creatine, and phosphatidylethanolamine to yield phosphatidylcholine.

117. The answer is D. *(White, ed 5. p 719.)* The final step of purine catabolism in primates is the oxidation of xanthine to uric acid. Overproduction or undersecretion of uric acid causes the urate crystal formation of gout. Urea and allantoic acid are the excreted products of fish. Xanthine and orotic acid are intermediates in pyrimidine metabolism which are not normally excreted.

118. The answer is A. *(White, ed 5. p 707.)* Purine synthesis is a metabolic pathway that is common to bacteria and man. Nitrogen fixation, photophosphorylation, mucopeptide synthesis, and fermentation occur in plants or bacteria but not in man.

119. The answer is C. *(White, ed 5. pp 706-707.)* Carbons 4 and 5 and nitrogen 7 of purines are derived from glycine. The glycine molecule is incorporated intact into the purine ring. Carbons 2 and 8 arise from formate. Carbon 6 arises from carbon dioxide.

120. The answer is E. *(Mahler, ed 2. pp 827-828.)* Glutamine (nitrogen-"B") condenses with CO_2 (carbon-"C") to form carbamoyl phosphate that reacts with amino group (nitrogen-"D") of aspartic acid to include the side chain carboxyl (carbon-"A") in the pyrimidine ring.

121. The answer is C. *(Harper, ed 15. p 385.)* An early step in pyrimidine synthesis is the reaction catalyzed by aspartate transcarbamoylase: carbamoyl phosphate + aspartate → carbamoyl aspartate. Carbamoyl phosphate is not an intermediate in purine biosynthesis. Features A, B, D, and E are common to the biosynthetic pathways of purines and pyrimidines.

122. The answer is D. *(White, ed 5. p 648.)* The mitochondrial carbamoyl phosphate synthetase is thought to be involved in urea biosynthesis in the liver, while the cytoplasmic enzyme is thought to be involved in pyrimidine biosynthesis. The mitochondrial enzyme is present in relatively high activity and is activated by acetylglutamate. The cytoplasmic enzyme is present in low activity and is inhibited by UTP.

123. The answer is D. *(White, ed 5. pp 245-246, 720-721.)* In bacteria, pyrimidines (CTP) lead to feedback inhibition of aspartate transcarbamoylase. Aspartate transcarbamoylase has been shown to consist of 12 subunits; six each of catalytic and regulatory subunits. In mammals, UTP inhibits pyrimidine biosynthesis at the level of the cytoplasmic carbamoyl phosphate synthetase rather than at the level of aspartate transcarbamoylase.

124. The answer is B. *(White, ed 5. p 199.)* Bovine pancreatic ribonuclease gives 3'-nucleoside monophosphates as products from RNA. It preferentially cleaves after phosphates that are attached to the 3'-hydroxyl of a pyrimidine nucleotide. Snake venom RNase preferentially gives nucleotide 5'-phosphates.

125. The answer is C. *(Mahler, ed 2. pp 217-220.)* $s^\circ = MD(1-\bar{v}p)$ RT where s° is the sedimentation coefficient, M is the molecular weight, $\bar{v}$ is partial specific volume, and p is the density, i.e., mass/volume, of the solvent. D, the diffusion coefficient, is determined in part by the molecular shape. The optical density has nothing to do with the sedimentation coefficient, but measures the light absorbance of a solution.

126. The answer is B. *(Mahler, ed 2. pp 235-241.)* The melting temperature of DNA is defined as the temperature at which half of the base pairs have been disrupted, i.e., half of the DNA has been denatured. Among the ways of assaying the denaturation is to measure the optical density of the DNA (usually at 260 nm). The hypochromicity of native DNA results in an optical density approximately 40 percent greater in denatured DNA.

127. The answer is D. *(Mahler, ed 2. pp 238-239.)* The melting temperature of duplex DNA is the temperature at which 50 percent of base pairs are denatured. Because C-G base pairs have three rather than two hydrogen bonds, a high content of C + G increases the melting temperature. According to Chargaff's rules, the content of A + G = C + T in duplex DNA.

128. The answer is B. *(White, ed 5. pp 192-197.)* The Watson-Crick model predicts a double-stranded helix with hydrogen bonds between A-T and G-C bases on the inside of the helix. The chains run in an antiparallel, or opposite direction. Covalent phosphodiester bonds exist only between sugar moieties of the main chain and not between their attached bases.

129. The answer is B. *(Mahler, ed 2. pp 256-258. White, ed 5. p 198.)* Histones are a family of basic proteins (rich in lysine and arginine) that are bound non-covalently but stoichiometrically to DNA. They are of low molecular weight and are separable into five classes, largely on the basis of lysine and arginine content. In comparing histone IV from calf thymus and pea buds, a remarkable degree of homology was noted: only 2 out of 102 amino acids differed in the proteins from such diverse organisms.

130. The answer is A. *(White, ed 5. p 192.)* During DNA replication thymine pairs with adenine and guanine with cytosine. The strands are synthesized in an antiparallel direction, i.e., the $5' \to 3'$ sequence mentioned specifies its complement in a $3' \to 5'$ direction.

131. The answer is C. *(White, ed 5. pp 760-765.)* AUG is the codon for both methionine and N-formyl methionine. In *E. coli*, AUG is the chain-initiating codon and N-formyl methionine is the first amino acid incorporated into the nascent polypeptide. In mammals, AUG is also thought to be the codon for chain-initiation, but methionine (rather than N-formyl methionine) is the N-terminal amino acid.

132. The answer is C. *(Harper, ed 15. pp 63-71.)* Because genes must be constant in all cells of one organism, the ratios of nucleotide bases (A/T: G/C) must be constant in these cells. These ratios may vary from species to species but should not vary with age or environmental factors. The differentiation of one tissue from another, in any single animal, depends upon the variable expression of a genetic code common to all the tissues of that animal.

133. The answer is A. *(White, ed 5. p 739.)* Semiconservative double-stranded DNA replication demands that from each half (strand) of the labelled parental DNA, a complement will be synthesized that maintains the parental structure intact. Therefore, the first round of replication in a cold (unlabelled) solution will yield two molecules that are half-labelled. The second round will yield two half-labelled and two unlabelled molecules of double-stranded DNA.

134. The answer is E. *(White, ed 5. pp 768-769.)* Insertion of one extra base causes a frame-shift and misreading of all the DNA beyond the mutation. All the other mutations cited cause misreading only at the site of that mutation or the loss of a single codon, and are, therefore, less deleterious to the message being transcribed.

135. The answer is B. *(White, ed 5. p 767.)* Dimerization between thymines on the same polynucleotide strand is caused by ultraviolet radiation (260 nm) and stops replication at that point. The dimers can be excised and repaired by enzymes that include ligase, or can be dissociated by further exposure to longer wavelengths (330-450 nm) or shorter wavelengths (230 nm) in the process of photoreactivation.

136. The answer is E. *(White, ed 5. pp 762, 767-769, 868.)* In Hb S a valine residue replaces a glutamic acid on the β chain, as the result of a point mutation in one nucleotide base. This single nucleotide alteration at the second position of the triplet consists in a change of thymine to adenine.

137. The answer is D. *(Mahler, ed 2. pp 940-949. White, ed 5. pp 764-765.)* The anticodon CGU is complementary to the codon ACG. Note that both triplets are written with the 5'-end at the left and the 3'-end at the right: Watson-Crick base pairing is antiparallel. The anticodon site in tRNA frequently contains unusual nucleotides, such as inosinic acid or pseudouridylic acid.

138. The answer is C. *(White, ed 5. pp 769-770.)* N-Formyl methionine is thought to be always the first amino acid at the N-terminus of polypeptides synthesized in *E. coli*. Two discrete tRNAs for methionine exist in *E. coli* to specify whether or not N-formylation is to occur.

139. The answer is A. *(Watson, ed 2. pp 411-414.)* According to the wobble hypothesis of Crick, the base pairing at the third position of the codon is less stringent than at the other two positions. Thus, certain bases at the 5'-position of the anticodon are able to pair with more than one base at the 3'-position of the codon. This allows one tRNA species to read more than one codon.

140. The answer is C. *(White, ed 5. p 772.)* The following three codons are chain-terminating (nonsense) codons: UAA, UAG, and UGA. Termination of a polypeptide chain requires mRNA of one of these codons, and liberation of the tRNA bound to the C-terminal residue of the chain.

141. The answer is D. *(Watson, ed 2. pp 337-348.)* Linear RNA molecules are transcribed from one strand of duplex DNA designated the sense strand. RNA polymerase synthesizes the RNA by adding ribonucleotide moieties to the growing 3′-end, and this chain may be circular.

142. The answer is A. *(Mahler, ed 2. pp 905-907.)* Sigma-factor is the subunit of RNA polymerase which confers specificity of initiation to the core enzyme. In the presence of sigma-factor, RNA polymerase will choose the correct strand of duplex DNA to transcribe, as well as starting transcription at the appropriate promoter region.

143. The answer is C. *(Harper, ed 15. p 66.)* The first step in protein synthesis involves the activation of amino acids. ATP is utilized to form aminoacyl adenylates and pyrophosphate. The aminoacyl adenylate then reacts with the appropriate tRNA species to yield the aminoacyl-tRNA.

144. The answer is A. *(White, ed 5. pp 770-772.)* Two molecules of GTP are hydrolyzed in the process of forming one peptide bond by the ribosome. Ribosomal peptidyl transferase is located in the 50 S subunit. The three distinct soluble peptide elongation factors are the transfer factors Tu and Ts, and the protein G. The protein G possesses ribosome-dependent GTPase activity.

145. The answer is C. *(White, ed 5. p 774.)* Puromycin is an analog of the 3′-end of an aminoacyl-tRNA. It can occupy the A-site of the 50 S ribosomal subunit and accept the carboxyl group of the nascent polypeptide chain. Consequently, polypeptidyl puromycin is released from the ribosomal complex and can be analysed.

146. The answer is D. *(Harper, ed 15. p 68.)* Tetracycline inhibits the binding of aminoacyl-tRNA to mRNA by preventing combination of aminoacyl-tRNA with initiator sites on 30 S subunits of ribosomes.

147. The answer is A. *(Harper, ed 15. p 68. Mahler, ed 2. pp 956-957.)* Chloramphenicol interacts with the 50 S subunit of prokaryotic ribosomes to inhibit the process of chain elongation. Streptomycin also binds to the 50 S subunit of ribosomes resulting in decreased rates of protein synthesis, but additionally it induces misreading of mRNA codons.

148. The answer is C. *(Mahler, ed 2. pp 862-865.)* 5-Bromouracil is mutagenic because it may vacillate between two configuations. In the enol form, 5-bromouracil pairs with guanine: in the keto form, it pairs with adenine. Thus, the presence of 5-bromouracil in place of thymine may lead to the replacement of adenine by guanine in the complementary strand of DNA.

149. The answer is A. *(White, ed 5. p 750.)* During translation, mRNA codons dictate the amino acids to be synthesized into proteins. This is in contrast to transcription which yields mRNA complementary to a sequence of DNA.

150. The answer is C. *(White, ed 5. pp 202-203, 302-304, 752.)* The two subunits of ribosomes are composed of proteins and rRNA. Ribosomes are found in the cytoplasm, in mitochondria, and bound to the endoplasmic reticulum. Transcription refers to the synthesis of RNA complementary to a DNA template, and has nothing immediately to do with ribosomes.

151. The answer is D. *(J Mol Biol 75:515, 1973.)* Cordycepin (3'-deoxyadenosine) is phosphorylated into RNA in an analogous fashion to adenosine. However, the analog terminates the growing nucleic acid polymeric chain and prevents, therefore, the formation of the poly A sequences characteristic of mRNA.

152. The answer is D. *(White, ed 5. p 764, 769-771.)* The initiation but not the elongation of polypeptide chain synthesis (in *E. coli*) involves N-terminal N-formylmethionine incorporation as specified by codons AUG and GUG. Peptide chain elongation upon the ribosomal mRNA employs peptidyl transferase after codon-specific binding of aminoacyl tRNA, using energy from GTP and the factors G, Tu, and Ts.

153. The answer is C. *(Davis, ed 2. pp 107-108.)* The cell walls of grampositive bacteria exhibit a 200 to 800 Å-thick, electron-dense, outer layer consisting of N-acetylmuramic acid, N-acetylglucosamine, and D- and L-amino acid peptides. The enzyme lysozyme partially hydrolyses cell walls.

154. The answer is C. *(Jawetz, ed 11. pp 44, 51.)* Transformation is characterized by the uptake of soluble DNA released by a donor cell. Transformation takes place only in bacteria that can utilize the high molecular weight DNA of the medium. Although originally discovered in the pneumococcus, transformation also occurs in *Hemophilus, Bacillus,* and *Neisseria* species.

155. The answer is C. *(Davis, ed 2. pp 1113-1118.)* Phage λ (lambda) transfers only a restricted group of genes which are located near the prophage to *E. coli*. These are the gal (galactose utilization) and bio (biotin synthesis) regions.

156. The answer is A. *(Jawetz, ed 11. p 44.)* Transduction is the process in which a piece of donor chromosome is carried to the recipient by a temperate bacteriophage produced by the donor cell. There are two types of transduction. In the generalized type, the phage has an approximately equal chance of transducing any segment of the donor's chromosome. In restricted transduction, the phages carry only segments that are immediately adjacent to the prophage.

157. The answer is D. *(Jawetz, ed 11. pp 44-45.)* Chromosomal material can be transferred between bacterial cells by transformation, transduction, and conjugation. Large segments of chromosomes, however, are passed only through conjugation.

158. The answer is B. *(Davis, ed 2. p 984.)* Cycloheximide is an antibiotic that inhibits protein synthesis by blocking the tranfer of amino acids in eukaryotic cells, i.e., tissue cultures (HeLa), yeasts (*Cryptococcus, Saccharomyces*), and protozoans (plasmodia). It is toxic to the host and, therefore, is useful only diagnostically in cultures where specific microorganisms are to be suppressed so that molds can be cultured.

159. The answer is E. *(Davis, ed 2. pp 1174-1177.)* Interferon is a cell-specific protein produced by cells infected by a virus. Interferon acts both on the cell in which it is produced and on those exposed to it by eliciting production of an antiviral protein. It appears that all animal cells are capable of yielding some interferon response, but that cells of the recticuloendothelial system are the major source in an infected animal. Interferon production is dependent on transcription and translation.

160. The answer is A. *(Jawetz, ed 11. pp 39-40.)* Spontaneous point mutations are most likely to arise from tautomeric shifts of electrons in a purine or pyrimidine base. This shift causes an altered base-pairing during DNA replication. Several mutagenic agents such as 5-bromouracil and 2-aminopurine increase tautomerization of bases in DNA.

161. The answer is A. *(Davis, ed 2. pp 256-258.)* Frame-shift mutations which are caused by the insertion or deletion of a single base-pair, are induced by acridine derivatives. 5-Bromouracil produces transitional mutations through tautomerization and subsequent base-pairing with adenine instead of guanine. Azathioprine is converted into 6-mercapto-purine, a purine analog. EES produces transitions through alkylation of guanine.

162. The answer is B. *(Jawetz, ed 11. pp 39-40.)* Proflavine (an acridine dye) is a chemical mutagen that intercalates between stacked base pairs of DNA. It thus causes the insertion or deletion of a single nucleotide base in a replicating strand and shifts the reading frame of the message from that point on. Ethylmethanesulfonate is an alkylating agent; nitrous acid causes deamination of adenine, guanine, and cytosine; 5-bromouracil incorporates into DNA in place of thymine. Each of these changes can cause mutation but will not cause a frame-shift mutation as does proflavine.

163. The answer is E. *(Mahler, ed 2. pp 904-911, 961-964.)* Transcription of the lactose operon is initiated by binding of RNA polymerase (including sigma factor) to the promoter region, "p." The region designated "i" governs repressor synthesis, and repressor protein from "i" is the negative control exerted upon the operon at the site designated "o." The regions "z," "y," and "a" specify β-galactosidase, galactoside permease, and acetylase enzymes.

164. The answer is B. *(White, ed 5. p 749.)* The smallest unit of DNA capable of regulating the synthesis of a polypeptide is the cistron. A gene sequence under the coordinated control of a single operator is called an operon. Recon and muton refer to the postulated smallest units of recombination and mutation.

165. The answer is D. *(Lehninger, 1970. p 644.)* Mitochondria of eukaryotic cells contain four to six molecules of circular DNA, as shown. The nuclear DNA of eukaryotes and the *E. coli* DNA genome are much larger. Lysosomes do not contain DNA.

166. The answer is C. *(Miller, 1972. p 66.)* The Hfr chromosome transfer is mediated by an integrated F factor. Since this transfer begins at a site within the F factor, the location of the integration site determines the order of transfer.

167. The answer is A. *(Hayes, ed 2. p 48.)* Since double cross-overs between two genetic loci result in no apparent recombination of the two loci, the recombination frequency is always less than the sum of the intervening frequencies. In the absence of double cross-overs, the recombination and cross-over frequencies would exhibit a direct linear relationship.

168. The answer is A. *(Hayes, ed 2. p 55.)* Transformation involves the transmission of genetic markers by fragments of bacterial DNA, independent of conjugation. The proportion of a whole bacterial chromosome that mediates transformation is usually less than one percent. This process does not occur in any predictable (linear) sequence.

169. The answer is D. *(Miller, 1972. p 64.)* Only the F pili have been associated with sexual activity. The transfer of the bacterial chromosome, the formation of recombinants, and the other properties mentioned do not necessarily occur during conjugation.

170. The answer is B. *(Miller, 1972. p 82.)* The streptomycin will kill all of the male cells, and only those female cells which have received the F' factor will be able to utilize the lactose.

171. The answer is B. *(Lewin, 1974. p 15.)* In a fully overlapping genetic code, a single base mutation would alter three amino acids: in a non-overlapping code only one amino acid is altered. With either coding system, more than one triplet might specify a given amino acid. Ribosomal binding and termination triplets have no bearing upon deciphering of the coding system. The fact that poly(U-G) directs the synthesis of poly(cys-val) is ambiguous in the deciphering of genetic coding.

172. The answer is C. *(Lewin, 1974. p 5.)* The major implication of the *cis/trans* test is that when two mutant sites lie on the same gene, they show complementation to give a wild phenotype in only the *cis* arrangement. Two mutant sites in different genes complement in either the *cis* or *trans* arrangement.

173. The answer is E. *(Lewin, 1974. p 51.)* Puromycin is a structural analog of the aminoacyl end of the tRNA. It irreversibly reacts with the peptidyl-tRNA, thereby terminating protein synthesis. Streptomycin, like tetracycline and chloramphenicol, inhibits ribosomal activity. Mitomycin covalently cross-links DNA, preventing cell replication. Rifampicin is an inhibitor of DNA-dependent RNA polymerase.

174. The answer is A. *(Lewin, 1974. p 505.)* In all cases tested, xeroderma pigmentosum appears to be due to the inability of an excision-repair system to remove thymine dimers, which are formed upon exposure of DNA to ultraviolet radiation. Mutagenesis by this mechanism is presumably the basis for the multiple neoplasms that occur in patients with this disease.

175. The answer is A. *(Miller, 1972. p 79.)* If an $F'lac^+$ factor is carried in a lac^- strain, cells which have lost the F' factor will not grow, as they must subsist in a lactose medium. The use of a $recA^-$ strain will prevent any incorporation of the $F'lac^+$ into the bacterial chromosome, as it is deficient in recombination.

176. The answer is D. *(Lewin, 1974. p 280.)* The term operon describes the system as a whole, including both structural genes and regulator elements. It is often linked physically in a cluster on the chromosome. Codons comprise three nucleic acids, and are the simplest unit of any genetic message. Cistrons are units of genetic analysis defined by recombination, and a muton is the smallest unit defined by mutation analysis. The genome is the entire genetic message of a species.

177. The answer is C. *(Lewin, 1974. pp 44, 76.)* One ATP is required for the aminoacylation of tRNA. The aminoacyl-tRNA requires one GTP for binding to the ribosome A site and a second GTP for translocation upon the ribosome.

178. The answer is D. *(Lewin, 1974. p 414.)* 5-Bromouracil incorporates into replicating DNA, in place of thymidine, to yield a heavier DNA. The newly synthesized DNA fragments can then be quantitated by centrifugation through density gradients of cesium chloride. 5-Bromouracil is not more reactive or sensitive to cleavage than thymidine, and it does not cause frame-shift mutations, as do the acridine dyes.

179. The answer is E. *(Miller, 1972. pp 73-74.)* The transfer of a prophage to a phage-sensitive recipient is called zygotic induction. The point of integration of the prophage is measured by the time from conjugation initiation to the phage burst in the recipient.

180. The answer is C. *(Miller, 1972. p 115.)* Acridines induce frameshifts in which the reading of the triplet code is advanced one nucleotide. The mechanism of this frame-shift appears to be intercalation of the dye between adjacent base pairs in the DNA double helix.

181. The answer is A. *(Lewin, 1974. p 262.)* The increased specificity of phage T4 synthesis in an infected cell is due to the substitution of phage-coded subunits into the host RNA polymerase.

182. The answer is C. *(Lewin, 1974. p 345.)* Lysogeny is maintained by the synthesis of viral regulator proteins which prevent the expression of phage genes needed for lytic development. The same regulators repress any other phage DNA of the same type and, therefore, confer immunity.

183. The answer is B. *(Lewin, 1974. p 310.)* The lac operon is shut off by a repressor protein that is inactivated by an inducer, i.e., negative control of induction.

184. The answer is A. *(Lewin, 1974. p 315.)* In positive control of repression, the operon is switched on by an inducer protein. The inducer protein can be inactivated by a co-repressor. The inducer protein itself requires no induction, derepression, or mediator.

185. The answer is E. *(Lewin, 1974. p 280.)* The repressor protein binds to a specific operator site on the operon. This prevents the RNA polymerase from transcribing the subsequent structural genes into mRNA. Repressor protein does not act distally in this mechanism, either to prevent translation or actual protein synthesis.

186. The answer is D. *(Hayes, ed 2. p 174.)* Hot spots are sites in DNA that are highly susceptible to the action of particular mutagens.

187. The answer is C. *(Hayes, ed 2. p 56.)* Transduction involves the use of bacterial viruses to transfer small regions of a bacterial chromosome from a donor to a recipient. It is of little use for genetic mapping; however, it can be a valuable method for the analysis of genetic fine structure.

188. The answer is D (4). *(Mahler, ed 2. pp 698-699. White, ed 5. pp 366-370.)* Oligomycin is an inhibitor of oxidative phosphorylation. By inhibiting the mitochondrial ATPase which synthesizes ATP, oligomycin prevents the use of the energy derived from electron transport to synthesize ATP. Oligomycin has no effect on coupling, but blocks mitochondrial phosphorylation so that both oxidation and phosphorylation cease in its presence.

189. The answer is E (all). *(White, ed 5. pp 770-774.)* GTP hydrolysis is the energy source for protein synthesis. ATP is necessary for charging tRNA with amino acids. Formation of new peptide bonds is catalyzed by the enzyme peptidyl transferase.

190. The answer is B (1, 3). *(White, ed 5. pp 188, 192.)* The sugar moiety of DNA is 2-deoxyribose where H has replaced the OH in the 2' position making it impossible to form a phosphate bond at this position. There are no phosphate → phosphate bonds in DNA (as exist in ADP and ATP). The main chain of the two antiparallel strands of a DNA molecule consists of deoxyribose residues linked by covalent phosphodiester linkages from 3'- to 5'-positions.

191. The answer is B (1, 3). *(White, ed 5. pp 732, 774.)* Methotrexate blocks the dihydrofolate reductase necessary to purine and pyrimidine synthesis. Actinomycin D blocks the DNA-dependent RNA polymerase. By complexing with guanine residues in DNA, it inhibits all RNA synthesis and thereby protein synthesis. Chloramphenicol and tetracycline are inhibitors of protein synthesis.

192. The answer is D (4). *(Jawetz, ed 11. pp 49-50.)* DNA transfer by Hfr donors occurs when a suspension of Hfr cells is mixed with an excess of F$^-$ cells. At random times, transfer will be interrupted by the spontaneous breakage of the DNA molecule. The Hfr state is reversible; detachment of the F factor takes place approximately once per 10^5 cells at each generation.

193. The answer is B (1, 3). *(Jawetz, ed 11. pp 45-50.)* Bacterial conjugation requires physical contact between donor and recipient cells joined by an F pilus. Small or large segments may be transferred, depending on the state of the F^+ factor. F^- cells can act only as recipients, while F^+ cells may behave as donors or recipients.

194. The answer is E (all). *(Jawetz, ed 11. pp 48, 51.)* Bacteria acquire drug resistance by spontaneous mutation, recombination, or transduction. Resistance to penicillin in gram-positive staphylococci is carried by transducing phages: penicillinase plasmids do not undergo conjugal transfer.

195. The answer is B (1, 3). *(Jawetz, ed 11. pp 105-106.)* Early proteins synthesized in response to phages include enzymes necessary for the synthesis of new phage DNA: DNA polymerase; thymidylate synthetase; and kinases for the formation of nucleoside triphosphates. The T2, T4, and T6 phages which incorporate hydroxymethylcytosine into their DNA cause the host cell to synthesize an additional series of enzymes needed for the synthesis of hydroxymethylcytosine.

196. The answer is A (1, 2, 3). *(Davis, ed 2. pp 189-190.)* High-frequency recombination (Hfr) strains of bacteria integrate the F agent in their chromosomes. In this position conjugation may occur but the F-attached chromosome is transferred rather than the F agent. All cells of the Hfr strain start chromosomal transfer at the same locus and produce an orderly sequence of gene entry and recombination which is higher in frequency for certain genes than would otherwise be predicted.

197. The answer is C (2, 4). *(Davis, ed 2. pp 120-121, 151.)* Bacitracin is a bacteriolytic cyclic peptide that blocks cell wall synthesis by inhibiting phosphorylation of undecaprenol-PP formed during the transfer of subunits from carrier to cell wall. Penicillin blocks cross-linking of the peptidoglycan (transpeptidation reaction). Chloramphenicol inhibits protein synthesis, while sulfonamides are competitive analogs of p-aminobenzoate.

198. The answer is E (all). *(Jawetz, ed 11. pp 45-46, 48.)* Bacteria contain extrachromosomal genetic elements known as plasmids which can mediate chromosome transfer via conjugation. The most widely studied plasmid is F, the sex factor of *E. coli* K12. Bacteria containing the sex factor may act as females or males and are able to conjugate with males or females.

199. The answer is C (2, 4). *(Jawetz, ed 11. p 127.)* Chloramphenicol, a strong inhibitor of protein synthesis, is effective against many gram-negative rods, cocci, and rickettsiae. It blocks amino acid attachment to nascent peptide chains by interfering with the action of peptidyl transferase.

200. The answer is A (1, 2, 3). *(White, ed 5. pp 197-205.)* Viruses can reproduce only inside living cells and are incapable of independent DNA replication. Prokaryotes (bacteria), as well as the nuclei and mitochondria of eukaryotic cells, are able to sustain DNA replication.

201. The answer is E (all). *(Harper, ed 15. p 68.)* All of the compounds mentioned bind to ribosomes to block peptide synthesis. The modes of action vary in detail. Tetracycline inhibits binding of aminoacyl tRNA to mRNA by interference at the level of the 30 S subunit of the ribosome. Streptomycin, by contrast, binds to the 50 S subunit of the ribosome, decreasing the rate of protein synthesis and also interfering with the reading of mRNA codons.

202. The answer is C (2, 4). *(Harper, ed 15. p 51.)* Thymine and deoxyribose are incorporated into DNA, not RNA. Adenine is incorporated into both RNA and DNA and uracil only into RNA.

203. The answer is A (1, 2, 3). *(Harper, ed 15. p 387.)* Purine ribonucleotide diphosphates may be reduced by NADPH to yield purine deoxyribonucleotide diphosphates. Ribonucleotide reductases are the system that accomplishes this reaction. The enzymes are cobamide-dependent, since it is a B_{12} containing coenzyme (5, 6-dimethylbenzimidazole) which functions as the hydrogen transferring agent. They accomplish reduction without glycoside cleavage, and are not unique for each nucleoside.

204. The answer is A (1, 2, 3). *(Harper, ed 15. pp 380-385.)* The ring nitrogens of purines are derived from glutamine (N-3, N-9), glycine (N-7), and aspartate (N-1).

205. The answer is A (1, 2, 3). *(Harper, ed 15. pp 382-383, 390-391.)* The rate-limiting step in the biosynthesis of purine nucleotides is thought to be the synthesis of 5-phosphoribosylamine from 5-phosphoribosylpyrophosphate and glutamine. This reaction is catalyzed by an enzyme which is subject to feedback inhibition by nucleotides of adenine and guanine, ATP, ADP, GDP, GMP, and also by inosine monophosphate. It has been proposed that desensitization of this enzyme to normal regulatory influences can lead to overproduction of uric acid and, as a result, to gout.

206. The answer is A (1, 2, 3). *(Harper, ed 15. p 387.)* Xanthine oxidase catalyzes the conversion of hypoxanthine or xanthine to uric acid. Purine nucleotides, e.g., AMP, GMP, IMP, are catabolized to uric acid via xanthine or hypoxanthine.

207. The answer is A (1, 2, 3). *(Lehninger, 1970. pp 210, 261.)* Cell membranes are composed of lipid and protein and do not include DNA. The translational complex refers to the combination of mRNA, tRNA, and ribosomes involved in protein synthesis.

208. The answer is D (4). *(White, ed 5. pp 194-196.)* G-C base pair bonds are stronger than adenine (A)-thymine (T) bonds. DNA with a high G-C content is more stable and melts at a higher temperature than DNA that is rich in A-T. The stability relates to the existence of two hydrogen bonds between A and T, compared to three between G and C.

209. The answer is D (4). *(Mahler, ed 2. pp 883-913.)* Both RNA and DNA polymerases share the property that bases are added at the free $3'$-end of the growing polynucleotide chain. DNA polymerase synthesizes DNA that is complementary to DNA. The enzyme that synthesizes DNA that is complementary to RNA is called reverse transcriptase.

210. The answer is D (4). *(Watson, ed 2. pp 349-350.)* Sigma factor functions exclusively during the initiation phase of transcription of RNA from a DNA template. During elongation, sigma factor is dissociated from the core enzyme. DNA-dependent RNA polymerase is neither simpler than DNA polymerase, nor does it require less than four nucleotide precursors. RNA synthesis is directed in a $5'$ to $3'$ direction.

211. The answer is A (1, 2, 3). *(Harper, ed 15. pp 69-71. Watson, ed 2. p 444.)* The collection of adjacent nucleotides that code for a single mRNA molecule under the control of a single promoter is called an operon. According to this definition, operons contain the sequence of adjacent nucleotides which code for a single continuous messenger RNA molecule. An operon may thus represent single, or multiple genes, which are controlled in turn by an adjacent operator. The repressor, on the other hand, is often coded for by remote portions of the genome, and may in fact be active upon more than one operon.

212. The answer is C (2, 4). *(Harper, ed 15. pp 69-70.)* Operator genes are the loci for binding of repressor or inducer molecules. According to whether or not the operator gene locus is occupied, the structural genes either will or will not be transcribed.

213. The answer is E (all). *(Lehninger, 1970. pp 716-718.)* Messenger RNA molecules code for one polypeptide chain. There are no "spacers" to mark the end of one codon and the beginning of another. There are three nonsense codons. According to the wobble hypothesis of Crick, certain anticodons can pair with more than one codon where variability occurs at the $3'$-position of the codon.

214. The answer is C (2, 4). *(Mahler, ed 2. pp 196-197.)* RNA is synthesized with only one of the four ribonucleotides, labeled at the alpha position with ^{32}P, for "nearest neighbor" analysis. Since alkaline hydrolysis gives $3'$-nucleotide monophosphates as products, alkaline hydrolysis allows determination of the frequency with which the $3'$-hydroxyl groups of various nucleotides are attached via phosphodiester bonds to the $5'$-hydroxyl group of the labeled precursor.

215. The answer is B (1, 3). *(Mahler, ed 2. pp 202-209.)* The correct Watson-Crick base pairings for DNA are adenine-thymine and cytosine-guanine. In RNA, uracil replaces thymine.

216. The answer is E (all). *(White, ed 5. pp 762-767.)* The triplet code is degenerate; there is more than one triplet that designates each amino acid. Often the degeneracy involves only the third base in the codon, which is less specific than the first two. Chain termination is determined by three codons: UAA; UAG; and UGA.

217. The answer is A (1, 2, 3). *(Davis, ed 2. pp 175-176. Hayes, ed 2. pp 56-60.)* Cell fusion allows for genetic recombination and is part of fertilization and conjugation. It occurs by various mechanisms in different forms of life. In haploid bacteria (such as *E. coli*) it occurs as a means of genetic transfer between male and female organisms. In higher forms, where the organism is diploid, miosis yields haploid ova and sperm that fuse by a more complex mechanism.

218. The answer is D (4). *(White, ed 5. pp 769-774.)* A growing peptide chain is attached to the free 3'-hydroxyl of AMP on the tRNA, and synthesis is begun at the amino terminal end.

219-225. The answers are: 219-A, 220-C, 221-D, 222-B, 223-B, 224-D, 225-B. *(Wintrobe, ed 7. pp 330-331, 333.)* Huntington's chorea is transmitted as an autosomal dominant disorder, while phenylketonuria, cystic fibrosis, and Wilson's disease are transmitted as autosomal recessive disorders. Vitamin D-resistant rickets is usually transmitted as an X-linked dominant disorder, although it can also be transmitted through autosomal dominant and recessive inheritance. Duchenne's muscular dystrophy and red-green color blindness are transmitted as X-linked recessive characteristics.

226-229. The answers are: 226-E, 227-B, 228-A, 229-B. *(Mahler, ed 2. pp 914-949.)* Messenger RNA is complementary to the base sequence of the DNA in an operon and codes for the polypeptide chains that correspond to the genes of the operon. Transfer RNA carries the anticodon triplet and is also covalently linked to the amino acid that corresponds to that anticodon. 16 S and 23 S rRNA's form parts of the 30 S and 50 S ribosomal subunits, respectively. The 30 S ribosome binds mRNA, and it is the 50 S ribosome that contains the peptidyl transferase enzyme activity.

Carbohydrates
and Lipids

---

230. The answer is C. *(White, ed 5. pp 418-513.)* Glyceraldehyde 3-phosphate dehydrogenase functions in both glycolysis and gluconeogenesis. Pyruvate carboxylase and fructose 1,6-diphosphatase are gluconeogenic enzymes. Pyruvate kinase and hexokinase are glycolytic enzymes.

231. The answer is A. *(Harper, ed 15. p 5.)* The molecule depicted in the question is α-D-glucose. It is one of a series of D-glucose hemiacetals in which there are alternate α and β forms available because of the asymmetry in the terminal carbon.

232. The answer is A. *(White, ed 5. pp 445-454.)* Gluconeogenesis is defined as the biosynthesis of free glucose, and applies to the resynthesis of glucose from lactic acid, and other metabolic intermediates, indirectly. It is the reverse of glycolysis.

233. The answer is D. *(Harper, ed 15. pp 262-264.)* The formation of glucose and carbohydrate from amino acids is called gluconeogenesis. The utilization of pathways of glycolysis or glycogenesis, and gluconeogenesis or protein synthesis, is governed by hormones such as epinephrine, the glucocorticoids, and glucagon.

234. The answer is A. *(Harper, ed 15. p 10.)* Hydrolysis of maltose yields glucose as the only product, as maltose comprises a disaccharide of α-D-glucopyranose. Glucose is combined with fructose in sucrose, and with galactose in lactose.

235. The answer is C. *(Lehninger, 1970. p 328.)* In glycogenolysis, phosphorylase catalyzes the synthesis of glucose 1-phosphate from glycogen and inorganic phosphate. Phosphoglucomutase converts glucose 1-phosphate to glucose 6-phosphate. Glucose 6-phosphate is a central compound at the junction of glycolytic, gluconeogenic, hexose monophosphate shunt, and glycogenic pathways. Glucose 6-phosphate may give rise to fructose 6-phosphate, and the Embden-Meyerhof pathway of glycolysis.

236. The answer is C. *(Harper, ed 15. pp 244-245.)* Aldolase catalyzes the reversible reaction fructose 1,6-diphosphate ⇌ glyceraldehyde 3-phosphate + dihydroxyacetone. This reaction in the Embden-Meyerhof pathway is catalysed by different aldolases in different tissues. ("A" in most tissues, and "B" in liver and kidney.)

237. The answer is E. *(Harper, ed 15. pp 243-248.)* ATP is synthesized by two reactions in glycolysis: the reaction catalyzed by phosphoglycerate kinase and that by pyruvate kinase. The second of these reactions is represented by G-H in the diagram for this question. In the diagram: A is glucose 6-phosphate; B is fructose 6-phosphate; C is fructose 1,6-diphosphate; D is glyceraldehyde 3-phosphate; E is 1,3-diphosphoglycerate; F is 3-phosphoglycerate; G is phosphoenolpyruvate; H is pyruvate.

238. The answer is D. *(Harper, ed 15. pp 297-299. Mahler, ed 2. pp 634-636.)* Prostaglandin (PG) $F_{1\alpha}$ is biosynthesized by a dioxygenase from homo-γ-linolenic acid. The prostaglandins as a whole are derived from eicosanoic acids which contain 20 carbons and three, four, or five double bonds. The PGE series has a keto group, and the PGF series has a hydroxyl group, in position 9.

239. The answer is E. *(White, ed 5. pp 566-567.)* Eicosa-5,8,11-trienoic acid is a member of the oleic acid family and can be synthesized, therefore, in the mammal biologically de novo. It is this acid that is observed to accumulate during deficiency of essential fatty acids, rather than the anabolites of linoleic and linolenic acids, and their family members, which comprise the remainder of the answers listed.

240. The answer is A. *(White, ed 5. pp 566-567.)* Eicosa-5,8,11-trienoic acid is part of the oleic acid family of polyunsaturated fatty acids and, therefore, can be synthesized de novo in higher animals. Double bonds in this family of fatty acids lie between the seventh carbon from the terminal methyl group and the carboxyl group. Of the remaining possibilities listed, dietary linoleic acid is required for the synthesis of arachidonic acid, and dietary linolenic acid is required for the synthesis of docosahexenoic acid. These four polyunsaturated fatty acids have double bonds within the terminal seven carbon atoms in common, and can not be made de novo.

241. The answer is B. *(Mahler, ed 2. p 733.)* Diglyceride + CDP-choline $\rightarrow$ CMP + phosphatidylcholine. Choline is initially phosphorylated by a kinase with ATP, and then reacts with cytidine triphosphate (CTP) to yield CDP-choline. CDP-choline is able to react with diglycerides to yield phosphatidylcholine, the most common phosphatide.

242. The answer is C. *(White, ed 5. pp 81-82.)* Cholesterol is a steroid. This class of compounds is derived from cyclopentaphenanthrenes, comprising three fused cyclohexane rings, and a terminal cyclopentane ring. Steroids with 8 to 10 carbon atoms at position 17, and an alcoholic hydroxyl group at position 3, are classed as sterols, of which cholesterol is the chief representative member in animal tissues.

243. The answer is A. *(White, ed 5. pp 590-595.)* Cholesterol is synthesized from acetyl CoA via 3-hydroxy-3-methylglutaryl CoA. All carbon atoms of endogenous cholesterol are derived from CoA immediately. Of course, any substance which is a precursor of acetyl CoA (e.g., pyruvate), may be a precursor of cholesterol.

244. The answer is D. *(White, ed 5. pp 598-599.)* Regulation of the biosynthesis of cholesterol is exerted by definition, at the "committed" and rate-controlling step. This is the reaction catalyzed by 3-hydroxy-3-methylglutaryl CoA reductase. This enzyme is reduced in activity by fasting, and by cholesterol feeding, thus exhibiting feedback control.

245. The answer is C. *(White, ed 5. pp 590-594.)* Squalene cyclizes to form lanosterol directly, which in turn is converted to cholesterol, via desmosterol. Squalene cyclization into a tetracyclic steroidal compound employs first, an epoxidase and activated molecular oxygen, and then an anaerobic cyclase, which yields lanosterol exclusively in mammals.

246. The answer is D. *(White, ed 5. pp 80, 597, 1190-1192.)* 7-Dehydrocholesterol is converted in the skin to cholecalciferol (vitamin D_3). Cholesterol is a sterol. The unrelated vitamin A is a carotenoid, and vitamin E is an aromatic compound whose side chain is similar to that of the terpenes.

247. The answer is C. *(Wintrobe, ed 7. p 1463.)* Because of the impaired absorption of fats and fat-soluble vitamins in malabsorption syndromes, the levels of serum carotene can be depressed. Carotenoids are precursors of vitamin A.

248. The answer is B. *(White, ed 5. pp 70-76.)* Sphingomyelin is an example of a phospholipid. Glycogen is a storage form of carbohydrate, while oleic acid is a fatty acid, and prostaglandin is a fatty acid derivative.

249. The answer is B. *(White, ed 5. pp 75-76.)* Cephalins, lecithins, sphingomyelins and plasmalogens are all phospholipids. Cerebrosides are glycolipids which lack phosphorus. Three important glycosphingolipids are the cerebrosides, the gangliosides, and the ceramide oligosaccharides. Cerebrosides are most abundant in the myelin sheath of nerves.

250. The answer is A. *(White, ed 5. pp 354-355.)* Cardiolipin is found almost exclusively in mitochondrial membranes. Although it is antigenic, antibodies prepared against isolated cardiolipins fail to react with intact membrane preparations from cells, suggesting that cardiolipin is buried within the matrix of the membrane.

251. The answer is A. *(White, ed 5. pp 587-588.)* Sphingosine, and not glycerol, constitutes the backbone of ganglioside molecules. Glycerol forms the backbone of glycerides, phosphatides, etc. Sphingosine is a long-chain aliphatic base which comprises the backbone of sphingomyelin and glycosphingolipids.

252. The answer is A. *(Mahler, ed 2. p 591.)* Chylomicrons, which have a triglyceride content in the range of 79 to 95 percent, have the lowest density among the substances listed. It is this property of chylomicrons that allows their facile detection on inspection of serum, in which they float after a period of 8 to 12 hours refrigeration.

253. The answer is D. *(White, ed 5. pp 393-394).* Cyanides inhibit the action of cytochrome oxidase, a key enzyme in the process of tissue respiration. In so doing, they reduce respiration by 60 to 90 percent, which indicates that this proportion of animal cell respiration involves the electron transport system terminating in cytochrome oxidase.

254. The answer is D. *(Davis, ed 2. p 136.)* Group translocation is the proper term for the phosphotransferase system's taking glucose from the external medium and releasing glucose 1-phosphate in the cell. This process differs from active transport, in which the compound, prior to and after transport across the membrane, is found in the same form.

255. The answer is C. *(Bondy, ed 7. pp 253-305.)* Blood-glucose levels in fasted diabetics are often low enough so as not to exceed the renal threshold. The symptomatic control of diabetes involves regulation so that blood glucose does not exceed renal thresholds and produce glycosuria.

256. The answer is C. *(McGilvery, 1970. pp 260, 296-299.)* Phosphorolysis of glycogen yields glucose 6-phosphate. One molecule of ATP is utilized in the synthesis of fructose diphosphate. Two molecules of ATP are formed at each of two steps: phosphoglycerate kinase; and pyruvate kinase. Thus, there is a net yield of three ATP molecules per glucosyl residue converted to lactate.

257. The answer is D. *(Harper, ed 15. pp 248, 257.)* Anaerobic glycolysis allows for the synthesis of two moles of ATP per mole of glucose. Although it is not known exactly how many moles of ATP may be synthesized by aerobic metabolism of glucose, it seems likely that this number is somewhere in the vicinity of 38 moles of ATP per mole of glucose.

258. The answer is A. *(Harper, ed 15. pp 243-257. Lehninger, 1970. pp 326-327.)* In glycolysis two molecules of glyceraldehyde 3-phosphate are oxidized by NAD, and two molecules of pyruvate are reduced by NADH, with no net oxidation-reduction change.

259. The answer is D. *(Mahler, ed 2. p 512. White, ed 5. pp 239-240.)* Fluoride is an inhibitor of enolase, an enzyme involved in the Embden-Meyerhof pathway of glycolysis. Fluoride combines with magnesium, calcium, and other metals in the enzymes it affects.

260. The answer is B. *(McGilvery, 1970. p 260.)* Aerobic glycolysis can be defined as the oxidative conversion of glucose to two molecules of pyruvate. In the process, two molecules of ATP and two molecules of NADH are produced. Since the two molecules of NADH produced in the cytoplasm must be transported into the mitochondrion for oxidation, it is not known how many ATP are produced. On the assumption that three ATP molecules are formed during the oxidation of one molecule of cytoplasmic NADH, the ATP yield of aerobic glycolysis can be calculated as eight ATP molecules per molecule of glucose.

261. The answer is C. *(Harper, ed 15. p 248.)* Fructose 1,6-diphosphate is synthesized from fructose 6-phosphate by phosphofructokinase. Aldolase catalyzes the reversible interconversion between fructose 1,6-diphosphate on the one hand, and glyceraldehyde 3-phosphate and dihydroxyacetone phosphate on the other hand.

262. The answer is D. *(Lehninger, 1970. pp 283-284.)* Glycolysis takes place in the cytoplasm. The remaining energy-associated functions occur in the mitochondria.

263. The answer is C. *(Lehninger, 1970. pp 300-303.)* High energy phosphates are defined in terms of an energy of hydrolysis greater than(more negative than) 7 kilocalories. All of the compounds listed, except glucose 6-phosphate ($\Delta F^{o\prime} = -3.3$ kilocalories) have a free energy of hydrolysis more negative than -7.0 kilocalories.

264. The answer is A. *(Lehninger, 1970. pp 290, 298-300.)* The terminal phosphate of ADP has a $\Delta F^{o} = -7.3$ kilocalories, about the same value as that for the hydrolysis of the terminal phosphate of ATP.

265. The answer is B. *(Mahler, ed 2. pp 605-630.)* A molecule of guanosine triphosphate is synthesized from guanosine diphosphate and phosphate at the cost of hydrolyzing succinyl CoA to succinate and CoA. This comprises substrate-level synthesis of high-energy phosphate.

266. The answer is B. *(Lehninger, 1970. pp 322-324.)* To calculate the overall change,

$$\Delta F^{o\prime} \text{ total} = \Delta F^{o\prime}_1 + \Delta F^{o\prime}_2 = +1.5 + -4.5 = -3.0$$

267. The answer is B. *(Harper, ed 15. p 256.)* The reaction catalyzed by α-ketoglutarate dehydrogenase is thought to be a thermodynamically irreversible reaction under physiologic conditions.

268. The answer is A. *(Lehninger, 1970. p 338.)* The step from citrate to isocitrate by way of *cis*-aconitate involves only the movement of a hydroxyl group and not donation of an H^+-e^- pair to the electron transport chain.

269. The answer is D. *(Harper, ed 15. p 256.)* Pyruvate dehydrogenase is not involved in the citric acid cycle. Fumarase, aconitase, isocitrate dehydrogenase, and succinate thiokinase are all part of the citric acid cycle. Pyruvate dehydrogenase is the enzyme complex that accomplishes oxidative decarboxylation of pyruvate to yield acetyl CoA in "active acetate."

270. The answer is C. *(Lehninger, 1970. p 276.)* Fats, sugars, and ATP are sources of chemical energy used by cells. Plants use sunlight as energy for photosynthesis. Living organisms cannot use heat as an energy source: they are isothermal and heat can do work only when it is transferred from a warmer to a cooler body.

271. The answer is B. *(Lehninger, 1970. pp 337-338, 486, 491.)* The intermediates α-ketoglutarate and oxaloacetate are formed by transamination of glutamic and aspartic acids. The citric acid cycle is an aerobic process providing the major catabolic pathway for glucose degradation and is a more efficient energy producing process than glycolysis.

272. The answer is B. *(White, ed 5. pp 342-351, 381.)* The regulation of the citric acid cycle remains a matter of controversy; there is no universal agreement on which enzymes are involved in its regulation. Among the choices listed in the question, pyruvate dehydrogenase is not involved in the citric acid cycle and there is no enzyme called citrate dehydrogenase. Malate dehydrogenase and aconitase are thought to catalyze quasi-equilibrium reactions. In short, isocitrate dehydrogenase is the only enzyme listed which is even a candidate for a role in regulation of the citric acid cycle.

273. The answer is E. *(White, ed 5. pp 414-415.)* In order of decreasing rate of absorption by the gastrointestinal tract monosaccharides may be ranked: galactose $>$ glucose $>$ fructose $>$ mannose $>$ xylose $>$ arabinose.

274. The answer is D. *(Mahler, ed 2. p 499.)* Glucokinase and hexokinase both catalyze the phosphorylation of glucose. While glucokinase is more specific for glucose than is hexokinase, the latter enzyme has a 1000-fold greater affinity for glucose. Only hexokinase is subject to feedback inhibition by glucose 6-phosphate. Glucokinase is present in liver but not in brain; hexokinase is present in both tissues.

275. The answer is D. *(McGilvery, 1970. pp 263-283.)* In the Cori cycle, glucose is converted to lactate in peripheral tissues (erythrocytes, renal medulla, etc.), and the lactate is returned to the liver for resynthesis into glucose.

276. The answer is B. *(Mahler, ed 2. p 474.)* Cellulose is a high molecular weight, D-glucose polymer, which yields the disaccharide cellobiose (4-O-β-D-glucopyranosyl-D-α-glucopyranose) on partial hydrolysis. Lactose is a disaccharide of glucose and galactose, whereas maltose is a homopolysaccharide of glucose, and sucrose is a disaccharide of fructose and glucose. Cellulose is a complex plant polysaccharide which is not digestible by humans.

277. The answer is E. *(McGilvery, 1970. pp 322-331.)* In a process such as fatty acid synthesis, the reducing equivalents (NADPH) are thought to derive from two principal sources: the hexose monophosphate shunt; and the malate cycle (the transhydrogenation pathway involving malate dehydrogenase and malic enzyme).

278. The answer is E. *(McGilvery, 1970. p 280.)* For each molecule of lactate, one ATP is utilized (either directly or indirectly) for the reactions catalyzed by pyruvate carboxylase, phosphoenolpyruvate carboxykinase, and phosphoglycerate kinase. Thus, six molecules of ATP are required for the synthesis of one molecule of glucose from lactate.

279. The answer is C. *(Lehninger, 1970. p 492.)* All of the compounds listed yield a net synthesis of glucose through the tricarboxylic acid cycle and glycolysis, except leucine which is converted into CO_2 or acetyl CoA. It is, therefore, spoken of as a purely ketogenic amino acid, as opposed to the others mentioned which are glycogenic.

280. The answer is E. *(Mahler, ed 2. pp 549-551.)* UDP glucose + glycogen$_n$ → UDP + glycogen $_{n+1}$
The nucleotide uridine diphosphate, in this and other ways, is closely related to carbohydrate metabolism.

281. The answer is A. *(Mahler, ed 2. pp 499-500.)* ATP + glucose → glucose 6-phosphate + ADP
ATP is the high-energy phosphate thus employed for phosphorylation of glucose via hexokinase or glucokinase. It is used also to phosphorylate fructose, and fructose 6-phosphate, but not glucose 1-phosphate which requires UTP.

282. The answer is C. *(Mahler, ed 2. p 520.)* Oxaloacetate + GTP → GDP + phosphoenolpyruvate + CO_2
Through this reaction's use of GTP, oxaloacetate may be decarboxylated and enter a pathway essentially the reverse of glycolysis.

283. The answer is D. *(Mahler, ed 2. p 478.)* Keratin is a protein. Chitin is composed of repeating N-acetyl-D-glucosamine residues, and both chondroitin sulfate and heparin contain sulfate groups.

284. The answer is E. *(Lehninger, 1970. p 236.)* Hyaluronic acid is present in the connective tissue ground substance of vertebrates, in synovial fluid, and in vitreous humor.

285. The answer is B. *(McGilvery, 1970. pp 579-582.)* $3'$-Phospho-adenosine-5'-phosphosulfate is the donor of sulfate in the biosynthesis of such sulfate esters as chondroitin sulfate.

286. The answer is C. *(White, ed 5. pp 68-69.)* Saponification is the name given to the alkaline hydrolysis of a neutral fat to yield a soap (salt of a fatty acid) and glycerol. It provides an initial analytic approach to neutral fats.

287. The answer is C. *(Lehninger, 1970. pp 513-516.)* The addition of CO_2 to acetyl CoA yields malonyl CoA, from which fatty acids are synthesized in two carbon increments to chain lengths of 18-20 carbon atoms. This occurs outside of mitochondria, using NADP rather than NAD.

288. The answer is E. *(White, ed 5. p 575.)* 3-Hydroxy-3-methylglutaryl CoA → acetoacetate + acetyl CoA
A major fate of acetoacetyl CoA in the liver is formation of 3-hydroxy-3-methylglutaryl CoA which is an intermediate in steroid and cholesterol synthesis. 3-Hydroxy-3-methylglutaryl CoA is then cleaved to free acetoacetate in the liver.

289. The answer is C. *(White, ed 5. pp 558-565.)* In the synthesis of fatty acids, acetyl CoA is synthesized within mitochondria and transported to the cytoplasmic compartment via citrate. Acetyl CoA carboxylase is thought to be the rate-limiting enzyme in the extramitochondrial portion of the lipogenic pathway. However, there is considerable evidence to suggest that the synthesis of acetyl CoA may be the rate-limiting step in the overall pathway of lipogenesis.

290. The answer is E. *(Harper, ed 15. pp 297-311.)* Thiokinase is involved in activation of fatty acids to CoA thioester derivatives. Acyl CoA dehydrogenase, enoyl hydrase, β-hydroxyacyl CoA dehydrogenase, and β-ketothiolase are all involved in the β-oxidation sequence, per se.

291. The answer is C. *(McGilvery, 1970. p 321.)* Acetyl CoA within the mitochondrion is condensed with oxaloacetate to form citrate. Citrate is then transported out of the mitochondrion. In the cytoplasm, citrate is cleaved to acetyl CoA and oxaloacetate at the cost of hydrolyzing ATP.

292. The answer is A. *(White, ed 5. pp 550-553.)* Fatty acid oxidation occurs in mitochondria by successive two-carbon unit shortening of the aliphatic chain to produce ATP. This process is stimulated by carnitine. The reactions are initiated by formation of the fatty acid thioester of CoA.

293.   The answer is C. *(Harper, ed 15. p 311.)* Thiokinase is the enzyme that catalyzes the ATP-requiring activation of fatty acids to CoA thioester derivatives. Fatty acyl CoA derivatives are used in many reactions, including triglyceride biosynthesis.

294.   The answer is B (1, 3). *(White, ed 5. p 43.)* In osazone formation, two molecules of phenylhydrazine react with the carbonyl function and the adjacent hydroxyl function in a sugar. In the cases of the commonly occurring hexoses, sugars which are identical in configuration at C-3, C-4, and C-5, e.g., glucose, mannose, and fructose, all give rise to the same osazones.

295.   The answer is C (2, 4). *(Mahler, ed 2. pp 466-472. White, ed 5. pp 486-487.)* Ribulose and fructose are ketose sugars. Ribose and glucose are aldoses. Isomerases allow the interconversion of aldose and ketose sugars, such as glucose and fructose, via phosphoglucose isomerase.

296.   The answer is E (all). *(White, ed 5. pp 325, 334, 339.)* Nucleotides such as ATP, NAD, and RNA all contain the sugar ribose. Acetyl CoA contains an ADP moiety which also contains ribose.

297.   The answer is C (2, 4). *(McGilvery, 1970. pp 272-280, 344, 356-357, 369-373, 376-381.)* Serine and glycerol are metabolized to pyruvate and dihydroxyacetone phosphate, respectively, both of which can be converted to glucose in mammalian liver. Oleic acid, which can be converted only to acetyl CoA but not to 3-carbon precursors of glucose, is not a substrate for gluconeogenesis. Similarly, leucine is not a precursor for synthesis of glucose.

298.   The answer is D (4). *(White, ed 5. pp 47, 49.)* Glycogen is a highly branched, energy-storage molecule which is hydrolyzed by α-amylase and β-amylase to yield, respectively, glucose and maltose (a disaccharide of glucose). Cellulose is a structurally important polymer of O-methylglucose.

299.   The answer is A (1, 2, 3). *(Mahler, ed 2. pp 477-478.)* Glycogen is a highly branched polysaccharide composed of α-D-glucose molecules linked in α-1,4 and α-1,6 glycosidic linkages. ·

300.   The answer is A (1, 2, 3). *(White, ed 5. pp 450-451.)* In gluconeogenesis, the reactions catalyzed by hexokinase and phosphofructokinase are effectively reversed by the enzymes glucose 6-phosphatase and fructose 1,6-diphosphatase, respectively. The pyruvate kinase reaction is reversed by the joint actions of pyruvate carboxylase and phosphoenolpyruvate carboxykinase.

301. The answer is B (1, 3). *(Mahler, ed 2. pp 549-550.)* Glycogen synthetase is an enzyme which transfers glucosyl moieties from UDP-glucose to a glycogen polymer primer. The enzyme exists in two forms: an active, dephosphorylated form; and an inactive, phosphorylated form. In plants, ADP-glucose plays a role similar to that of UDP-glucose in animals.

302. The answer is D (4). *(White, ed 5. pp 454-456.)* Glucose 6-phosphate dehydrogenase is the first enzyme specifically involved in the pentose cycle (hexose monophosphate shunt). Fumarase and α-ketoglutarate dehydrogenase are Krebs cycle enzymes. Hexokinase is the initiating enzyme responsible for phosphorylation of glucose in the Embden-Meyerhof glycolysis pathway.

303. The answer is E (all). *(Mahler, ed 2. pp 516-530, 783.)* Pyruvate is an important compound in intermediary metabolism. It is produced by glycolysis and may be oxidized via pyruvate dehydrogenase with the resulting acetyl CoA being oxidized by the citric acid cycle. Under some circumstances, pyruvate may serve as precursor for gluconeogenesis. In some bacteria, pyruvate reacts with aspartic semialdehyde in an early stage of lysine biosynthesis.

304. The answer is B (1, 3). *(Harper, ed 15. p 255.)* NAD and FAD are two of the cofactors involved in the oxidative decarboxylation of pyruvate by pyruvate dehydrogenase. Another is thiamine diphosphate. Neither folic acid nor NADP are involved in the decarboxylation of pyruvate.

305. The answer is E (all). *(Harper, ed 15. pp 254-255, 266-267.)* Arsenite is an inhibitor of lipoic acid-containing enzymes, e.g., α-ketoglutarate dehydrogenase. Malonate is an inhibitor of succinate dehydrogenase, and fluoroacetate can be converted to fluorocitrate which is an inhibitor of aconitase. The citric acid cycle requires oxygen and does not proceed anaerobically.

306. The answer is A (1, 2, 3). *(Mahler, ed 2. pp 626-629.)* The regulation of the citric acid cycle is only incompletely understood. However, some factors of regulatory importance have been implicated: The level of oxaloacetate may affect the citrate synthetase reaction directly; NAD^+ is a substrate for isocitrate dehydrogenase, 2-oxoglutarate dehydrogenase, and malate dehydrogenase; the ADP/ATP ratio may affect the citrate synthetase reaction, as well as the reaction catalyzed by succinate thiokinase.

307. The answer is B (1, 3). *(White, ed 5. pp 345-346.)* The α-ketoglutarate dehydrogenase complex employs the cofactors NAD^+, FAD, thiamine pyrophosphate, lipoic acid, and CoA. This enzyme yields succinyl-CoA in a reaction which is virtually unidirectional because of energetics favoring the product.

123

308. The answer is B (1, 3). *(Mahler, ed 2. pp 495-498.)* Glycolysis can proceed anaerobically with the production of lactate from glucose. The NADH produced at the triose phosphate dehydrogenase step is reoxidized by way of lactate dehydrogenase.

309. The answer is B (1, 3). *(White, ed 5. pp 427-429.)* Citrate and ATP are two allosteric inhibitors of phosphofructokinase. Cyclic AMP and NH_4^+ have both been reported to increase the activity of phosphofructokinase under certain conditions in vitro.

310. The answer is A (1, 2, 3). *(Mahler, ed 2. pp 692-694.)* According to the Mitchell hypothesis of oxidative phosphorylation, protons are actively transported out of mitochondria with the energy supplied by the electron transport chain. As protons re-enter mitochondria down an electrochemical gradient, they provide the energy for ATP synthesis. Uncouplers make membranes "leaky" to protons. It is thought that the intramitochondrial pH is higher than the extramitochondrial pH.

311. The answer is C (2, 4). *(Mahler, ed 2. p 685.)* The inner membrane of the mitochondrion is impermeable to NADH itself. In some species, i.e., insects, dihydroxyacetone phosphate is reduced to glycerol 3-phosphate which is oxidized by a flavoprotein bound to the inner membrane of the mitochondrion. In other species, i.e., mammals, it is thought that oxaloacetate is reduced to malate which is transported into the mitochondrion. Since the mitchondrial membrane is not permeable to oxaloacetate, it is necessary that the oxaloacetate be transaminated to aspartate for transport out of the mitochondrion.

312. The answer is B (1, 3). *(Mahler, ed 2. pp 500-503.)* Glycogen phosphorylase is subject to at least two modes of regulation. The enzyme exists in two forms: phosphorylase *b*, which is inactive and unphosphorylated, and phosphorylase *a*, which is active and phosphorylated. Phosphorylase *b* exhibits enzymatic activity in the presence of the allosteric activator AMP. Cyclic AMP leads to the conversion of the enzyme to the active, phosphorylated form, i.e., phosphorylase *a*.

313. The answer is C (2, 4). *(Mahler, ed 2. pp 35-40.)* "Energy-rich" compounds are those that have a free energy of hydrolysis that is large in absolute value and negative in sign. In some instances, (ATP), the availability of more resonance forms for the products of hydrolysis, (ADP and inorganic phosphate), contributes to the highly negative free energy of hydrolysis. Because of the ionization of the product, esters of amino acids are high energy compounds.

314. The answer is E (all). *(Mahler, ed 2. p 24.)* Hydrolysis of the phosphate in all of the compounds listed, including $ADP \rightarrow AMP + P_i$, yields a free energy change of more than -7 kilocalories/mole. Free energy changes above -5 kilocalories/mole in hydrolysis of phosphate bonds are considered to be "high."

315. The answer is C (2, 4). *(White, ed 5. pp 355-359.)* Oxidative phosphorylation can be uncoupled from respiration by chemicals like dinitrophenol without affecting the function of electron transport. $\Delta F'_0 = -52.7$ kilocalories for the transfer of a pair of electrons from NADH to O_2.

316. The answer is C (2, 4). *(Mahler, ed 2. pp 473-479.)* Hyaluronic acid and heparin are both mucopolysaccharides. Neuraminidase is an enzyme which is not composed of mucopolysaccharide, but of protein. Cellulose is a polysaccharide.

317. The answer is A (1, 2, 3). *(Lehninger, 1970. p 217.)* There are only one or two repeating units in most polysaccharides, and because of this simplicity they cannot function as informational macromolecules as do proteins and nucleic acids.

318. The answer is B (1, 3). *(White, ed 5. p 83.)* Bile acids are often conjugated with glycine to form glycocholic acid, and with taurine to form taurocholic acid. In human bile, glycocholic acid is found much more commonly.

319. The answer is A (1, 2, 3). *(White, ed 5. pp 59-87.)* Lipids are hydrophobic, water-insoluble substances that may contain phosphate and nitrogen in addition to carbon, hydrogen, and oxygen.

320. The answer is C (2, 4). *(Mahler, ed 2. pp 480-484.)* Mureins are composed, in part, of a central core which is an alternating heteropolymer of N-acetyl-D-glucosamine and N-acetylmuramic acid.

321. The answer is A (1, 2, 3). *(White, ed 5. pp 575-578, 1110.)* Starvation results in the increased use of lipids as an energy source, with increased fatty acid oxidation and production of acetoacetyl CoA, a precursor of ketone bodies. The utilization of ketone bodies for metabolism is also seen in insulin deficiency (diabetes mellitus), but not in diabetes insipidus. Diabetes insipidus results from a lack of vasopressin, and is manifest by polyuria and polydipsia.

322. The answer is C (2, 4). *(White, ed 5. pp 554-555, 558-565.)* Pyruvate and acetyl CoA are important intermediates in the conversion of glucose to fatty acids. The acetyl CoA exits from the mitochondrion in the form of citrate, not as acetyl carnitine. Acetyl carnitine occurs in the mitochondrial fatty acid oxidative process.

323. **The answer is D (4).** *(White, ed 5. p 566.)* Linoleic and linolenic fatty acids cannot be synthesized by mammals and are, therefore, referred to as essential. Other fatty acids, including palmitic, stearic, and oleic acids, do not contain double bonds between the seventh carbon from the terminal methyl group and the carboxyl group, and thus can be made by alternate desaturation and elongation in mammals.

324. **The answer is B (1, 3).** *(Mahler, ed 2. pp 592-599, 714-722.)* In lipogenesis, NADPH is the source of reducing equivalents: in fatty acid oxidation, both FAD and NAD^+ serve as cofactors. In lipogenesis, both the acyl carrier protein and CoA are pantetheine containing coenzymes, or prosthetic groups, involved in the pathway. In lipogenesis, acetyl CoA is carboxylated to yield malonyl CoA in order to activate the acetate moiety for addition to the growing fatty acid. The reactions of lipogenesis proceed extramitochondrially: fatty acids are oxidized within the mitochondria.

325. **The answer is E (all).** *(White, ed 5. pp 556-557.)* Propionic acid is activated to propionyl CoA which is carboxylated to S-methylmalonyl CoA. After racemization to R-methylmalonyl CoA, the molecule is rearranged by way of a vitamin B_{12}-dependent reaction to give succinyl CoA.

326. **The answer is C (2, 4).** *(Lindros KO: J Biol Chem. 249:7956-7963, 1974.)* The principal pathway for hepatic ethanol metabolism is thought to be oxidation to acetaldehyde in the cytoplasm by alcohol dehydrogenase. Acetaldehyde is then oxidized, probably within the mitochondrion, to yield acetyl CoA. Neither acetone nor methanol appear in this biodegradation pathway.

327. **The answer is A (1, 2, 3).** *(White, ed 5. pp 590-595.)* Cholesterol is the precursor of progesterone in a series of reactions leading from cholesterol to progesterone to the sex hormones and adrenal corticosteroids. Mevalonic acid is an early precursor, and squalene, lanosterol, zymosterol, and desmosterol, are all late precursors of cholesterol.

328. **The answer is C (2, 4).** *(White, ed 5. pp 1050-1057, 1062-1067.)* Cortisol and aldosterone have 11β-hydroxyl functions. Estradiol and testosterone lack an 11-OH function, but possess other hydroxyl functions which are individually specific.

329. **The answer is A (1, 2, 3).** *(Harper, ed 15. pp 474-479.)* Hydroxylation at the 17-position is a part of the biosynthetic pathways leading to estradiol, cortisol, and testosterone. Aldosterone is hydroxylated at the 18-position, and then desaturated to form an aldehyde.

330. The answer is C (2, 4). *(Harper, ed 15. pp 474-483.)* In general, C-21 hydroxylation is necessary for both glucocorticoid and mineralocorticoid activity. There is a 21-hydroxylase that is required to hydroxylate at C-21 in the biosynthesis of cortisol and aldosterone. Androgens, in general, lack carbons 20, 21 of the steroid nucleus, and are designated C-19 steroids for this reason. Estrogens are C-18 steroids.

331. The answer is E (all). *(White, ed 5. pp 70-75, 579-590.)* Phosphoglycerides contain glycerol phosphate, two fatty acid molecules, and either choline, ethanolamine, serine, or inositol bound in an ester linkage to phosphoric acid. Depending upon their constitution, they may be designated specifically phosphatidyl choline, phosphatidyl ethanolamine, phosphatidyl serine, and phosphoinositide.

332. The answer is E (all). *(Harper, ed 15. p 305. White, ed 5. pp 547-548.)* Chylomicrons contain triglycerides (79-95 percent), cholesterol (1-5 percent), phospholipids (3-15 percent), and protein (0.5-2.5 percent). They are found in chyle, which is the form in which fats are absorbed via the intestinal lymphatics.

333. The answer is D (4). *(Jawetz, ed 11. p 69.)* The glyoxylate cycle appears in microorganisms that are furnished only acetate as a source of carbon for growth, under aerobic conditions. Acetate is metabolized to succinate, and so provides the source of C_4 compounds that cannot, under these conditions, be provided by the tricarboxylic acid cycle.

334. The answer is C (2, 4). *(White, ed 5. pp 514-517.)* Biosynthesis of starch during photosynthesis occurs via the Calvin cycle (pentose phosphate pathway). CO_2 fixation in this reaction pathway has been demonstrated to occur in the dark, after light-priming of chloroplasts, yielding 3-phosphoglyceric acid from ribulose diphosphate and CO_2.

335. The answer is A (1, 2, 3). *(White, ed 5. p 522.)* Levan is produced by *Leuconostoc* by levansucrase, a transglycosylase, from sucrose. It is a fructose polymer similar to dextran, a glucose polymer.

336. The answer is C (2, 4). *(White, ed 5. pp 522-523.)* Dextran is a glucose homopolysaccharide which can be used in medicine as a colloid in place of whole blood or plasma.

337. The answer is C (2, 4). *(Davis, ed 2. p 117.)* Lipid A is the base on which the core polysaccharide forms. Ketodeoxyoctonic acid is considered to be part of the lipid A complex. The core polysaccharides are similar for members of each group of bacteria in the Enterobacteriaceae, and can be used to define them antigenically.

338. The answer is D (4). *(White, ed 5. pp 550-559.)* Beta-oxidation of fatty acids proceeds within mitochondria, and is the major fate of fatty acids. Fatty acid oxidation by other routes, i.e., α- and ω-oxidation, may proceed in the microsomal, soluble fraction of cells.

339. The answer is D (4). *(Harper, ed·15. p 109.)* Lipoic acid is one of the cofactors involved in the oxidative decarboxylation of α-ketoacids, e.g., pyruvate and α-ketoglutarate.

340. The answer is C (2, 4). *(Mahler, ed 2. pp 506, 716. White, ed 5. pp 426, 560.)* Citrate activates acetyl CoA carboxylase and inhibits phosphofructokinase. Phosphofructokinase is the enzyme which represents the "committed step" in the glycolytic sequence, while acetyl CoA carboxylase catalyses the rate-limiting step in fatty acid synthesis. The regulatory influence of citrate upon these enzymes is appropriate to optimize use of available substrate. Citrate and isocitrate influence acetyl CoA carboxylase equivalently.

341. The answer is D (4). *(Bondy, ed 7. pp 1133-1135.)* The Porter-Silber reaction requires hydroxyl functions at both C-17 and C-21, as well as a 20-keto function. Therefore, while tetrahydrocortisol is a "17-hydroxysteroid," cortol is not, despite the presence of a 17-hydroxyl function.

342. The answer is C (2, 4). *(Bondy, ed 7. pp 1133-1136.)* 17-Ketogenic steroids are those that can be converted to 17-ketosteroids by oxidation with sodium bismuthate. This reaction requires a hydroxyl group at the 17-position as well as at either C-20 or C-21. Thus, both cortol and tetrahydrocortisol, two degradation products of cortisol, are 17-ketogenic steroids.

343. The answer is B (1, 3). *(Bondy, ed 7. pp 1111-1120, 1559-1560.)* 11-Ketoetiocholanolone and androsterone are both 17-ketosteroids. The presence of an 11-keto function is suggestive of adrenal rather than gonadal origin of the steroid in question because the adrenal gland has an 11-hydroxylase enzyme which is absent from ovary and testis.

344. The answer is D (4). *(White, ed 5. pp 548-568.)* The mammalian polydesaturase enzyme is extramitochondrial and membrane-bound (sedimentable at 100,000 g for 60 minutes). Microsomal enzyme systems also exist for elongation of saturated and unsaturated fatty acyl CoA derivatives.

343-347. The answers are: 345-C, 346-D, 347-B. *(Mahler, ed 2. pp 472-474.)* Sucrose is α-D-glucopyranosyl-(1→2)-β-D-fructofuranoside. Lactose is β-D-galactopyranosyl-(1→4)-D-glucopyranose. Maltose is α-D-glucopyranosyl-(1→4)-D-glucopyranose.

348-349. The answers are: 348-E, 349-A. *(Mahler, ed 2. pp 473-487.)* Mureins are mucopolysaccharides which are the principal components of cell walls of both gram-negative and gram-positive bacteria. Keratan sulfate is a principal polysaccharide component of connective tissue. Starches are an intracellular polysaccharide storage form.

350-352. The answers are: 350-A, 351-E, 352-B. *(Harper, ed 15. p 248.)* ATP transfers its terminal phosphate to fructose 6-phosphate, producing fructose 1,6-diphosphate in a reaction which also requires Mg^{++}. Pyruvate is reduced to lactate by NADH in a reaction that has a large negative free energy change. Glyceraldehyde 3-phosphate is oxidized by NAD to the corresponding acid which is phosphorylated with inorganic phosphate to yield 1,3-diphosphoglycerate.

353-355. The answers are 353-A, 354-B, 355-C. *(Williams, ed 5. pp 854-867.)* Structurally the compounds shown should easily be identified as prostaglandins. If the structure is not sufficient for recognition, the facts that PGA is not active in stimulating nonvascular smooth muscle, and that PGF causes a transient increase in blood pressure, are sufficient for proper identification.

Vitamins and Hormones

356. The answer is E. *(Wintrobe, ed 7. p 602.)* Hartnup's disease consists of a pellagra-like skin rash following exposure to sunlight, intermittent cerebellar ataxia, psychosis, renal aminoaciduria, and the excretion of large amounts of indole 3-acetic acid and indican in the urine.

357. The answer is A. *(Beeson, ed 14. pp 59-60.)* Lead interferes with: 1) The combination of glycine and succinic acid to form delta-aminolevulinic acid (ALA); 2) The coupling of the two molecules of ALA to form porphobilinogen, by interfering with ALA synthetase and ALA dehydrase. ALA and coproporphyrin III accumulate and are excreted in the urine in excessive amounts. Normally, 2 mg or less of ALA are excreted in 24 hours: lead poisoning causes a 20 to 200 fold increase. This excretion increases before any other chemical or hematologic changes are manifest, and before the lead level is assayable in the blood.

358. The answer is B. *(White, ed 5. pp 220, 399.)* Tyrosinase catalyzes the oxidation of phenol derivatives to quinones, and contains the copper ion. Copper is also seen in electron transport enzymes, and in oxidases of ascorbate, and cysteine.

359. The answer is E. *(Williams, ed 5. p 536.)* Glucagon is known to cause increased lipolysis and mobilize energy stored in fatty tissues. Insulin, beta-adrenergic blockers, and nicotinic acid all lower the blood level of fatty acids, and decrease lipolysis. Prostaglandin E, in low dosages, increases lipolysis, while in larger doses it decreases lipolysis.

360. The answer is B. *(Williams, ed 5. p 544.)* After two days of fasting, liver and muscle glycogen supplies are essentially gone and amino acids become the prime precursors for glucose formation (gluconeogenesis). Most of the amino acids must then be taken from protein that is serving an important role in cell structure, enzyme activities, and other functions. A significant amount of alanine can no longer be derived from pyruvate. After prolonged starvation, gluconeogenesis decreases and the plasma alanine level is less than after an overnight fast. That this is a limiting factor in the amount of glucose produced by the liver is substantiated by the fact that infusion of alanine promptly increases the blood glucose level.

361. The answer is A. *(White, ed 5. pp 670-672, 1044-1047, 1102-1103, 1109-1111.)* Both vasopressin and oxytocin are secreted by the neurohypophysis. Vasopressin is a pressor and an antidiuretic and oxytocin stimulates uterine muscle contraction. The other pairs of antagonistic hormones cited refer to: B – glucose homeostasis; C – melanization; D – calcium metabolism; E – vasomotor tone.

362. The answer is C. *(Williams, ed 5. p 527.)* Insulin allows the disposition and utilization of glucose–particularly exogenous glucose. Gluconeogenesis is the adaptive response of the organism to low levels of glucose and is, therefore, diminished by insulin.

363. The answer is E. *(Williams, ed 5. pp 527-528.)* Among the potential uses for glucose, proteogenesis is as important as lipogenesis and glycogenesis. All of these are facilitated by insulin, not diminished.

364. The answer is D. *(Sawin, 1969. pp 32-42.)* A large protein, previously called the Van Dyke protein, now called neurophysin, binds oxytocin and vasopressin before they are released in the posterior pituitary.

365. The answer is D. *(Williams, ed 5. pp 91-92.)* Oxytocin plays only a secondary role in expulsion of the fetus. Its major role is in lactation. Suckling, via sensory pathways, in most mammals (but not necessarily in humans) causes oxytocin release, which in turn causes contraction of the mammary alveoli and ductules to force the milk into the larger collecting ducts and cisterns.

366. The answer is D. *(Williams, ed 5. pp 854-868.)* Although prostaglandins were originally isolated from prostate glands, seminal vesicles, and semen, their synthesis in other organs has been documented numerous times, and there are few organs in which prostaglandin release has not been documented.

367. The answer is D. *(Williams, ed 5. p 515.)* There are two phases of insulin release during glucose stimulation: an acute release which immediately follows glucose stimulation and depends on readily accessible presynthesized insulin for release; and a secondary phase which takes more time as it involves mobilization of stored granules of insulin.

368. The answer is B. *(Beeson, ed 14. pp 1368-1374.)* Riboflavin is a light-sensitive yellow compound that is the active coenzyme for several flavoproteins. Manifestations of riboflavin deficiency include corneal vascularization, angular stomatitis, cheilosis, nasolabial seborrhea, and scrotal, vaginal, and rectal dermatitis. Eye signs of thiamine deficiency are nystagmus and abducens palsy. Avitaminosis A produces nyctalopia and keratomalacia, and neither inositol nor niacin deficiency have characterisitic eye signs.

369. The answer is D. *(Lynch, ed 2. p 394.)* The glucose tolerance test illustrated shows the typical lag curve of delayed insulin response with a hypoglycemic period occurring in two to three hours, or later. This result may be associated with adiposity and early diabetes.

370. The answer is B. *(White, ed 5. pp 1094-1095, 1105-1131.)* Oxytocin and vasopressin are both nonapeptides that differ at only two positions: oxytocin has isoleucine and leucine at the 3- and 8-positions, respectively; vasopressin has phenylalanine and arginine at the 3- and 8-positions, respectively. Insulin differs from proinsulin in being much smaller. All the other possibilities cited are remote from each other.

371. The answer is C. *(White, ed 5. pp 1078-1093.)* The compound shown is cholesterol, one of a large group of steroids. Other members of the general group of steroids derived from cholesterol function as vitamins and hormones.

372. The answer is A. *(White, ed 5. pp 82-84, 1023, 1081.)* ACTH (adrenocorticotropin) is a peptide hormone of the adenohypophysis which influences the secretion of adrenocortical steroid hormones. The other compounds listed contain the basic steroid-ring system and are ultimate derivatives of cholesterol.

373. The answer is D. *(White, ed 5. pp 1105-1130.)* Vasopressin is elaborated by hypothalamic cells and transported to the posterior pituitary, or neurohypophysis for release. All of the other hormones cited are produced by the adenohypophysis, and released there.

374. The answer is C. *(White, ed 5. pp 865-867.)* Ferrous iron (Fe^{+2}) is the form absorbed in the intestine by ferritin, transported in plasma by transferrin, and stored in the liver in combination with ferritin, or as hemosiderin. There is no known excretory pathway for iron.

375. The answer is C. *(White, ed 5. pp 1138, 1144-1148.)* Certain amino acids and lipids are dietary necessities because humans cannot synthesize them. The energy usually obtained from carbohydrates can be obtained from lipids and the conversion of some amino acids to citric acid cycle intermediates.

376. The answer is E. *(Lehninger, 1970. pp 379-381, 421.)* Carnitine increases fatty acid transport into mitochondria and thus stimulates fatty acid oxidation. The other compounds listed are members of the electron transport chain.

377. The answer is C. *(White, ed 5. pp 1176-1179.)* Pernicious anemia results from an inability to absorb vitamin B_{12} from the gastrointestinal tract. This may be due to a deficiency of intrinsic factor, surgical gastrectomy, or distal ileal and small bowel disease. The earliest clinical signs of pernicious anemia follow the onset of vitamin B_{12} deficiency by 3 to 5 years.

378. The answer is D. *(Williams, ed 5. pp 754-755.)* Osteomalacia is the name given to the disease of bone seen in adults with vitamin D deficiency: it is analogous to rickets, seen in children with the same deficiency.

379. The answer is D. *(White, ed 5. pp 1182-1183.)* Scurvy results from ascorbic acid (vitamin C) deficiency. It results in failure of mesenchymal cells to form collagen, causing skeletal, dental, and connective tissue deterioration. Anemia in scurvy is associated with impaired utilization of iron, and deranged folate metabolism.

380. The answer is E. *(Harper, ed 15. p 255.)* Thiamine pyrophosphate is a coenzyme involved in oxidative decarboxylation of α-keto acids. Other cofactors involved in this type of reaction include FAD, NAD^+, lipoic acid, and CoA.

381. The answer is B. *(Mahler, ed 2. p 410.)* Pellagra is a disease resulting from deficiency of the vitamin niacin (nicotinic acid). The clinical syndrome characteristic of pellagra, which can include dermatitis, stomatitis, diarrhea, and dementia, may actually result from deficiencies of other nutrients in addition to niacin.

382. The answer is D. *(White, ed 5. pp 1154, 1160, 1173, 1182, 1190.)* Lack of vitamin D, a fat-soluble vitamin involved in calcium metabolism, produces rickets in children. Lack of the water-soluble vitamins: thiamine, niacin, B_{12}, and vitamin C, produces beriberi, pellagra, pernicious anemia, and scurvy, respectively.

383. The answer is C. *(White, ed 5. pp 1155-1156.)* Both beriberi and Wernicke's disease are thought to result from deficiency of thiamine (vitamin B_1). Birds manifest thiamine deficiency with opisthotonos, and both pigeons and foxes with thiamine deficiency suffer intraventricular hemorrhages and Chastek paralysis.

384. The answer is B. *(White, ed 5. p 456.)* The cofactors involved in the oxidative decarboxylation of α-keto acids include thiamine pyrophosphate, lipoic acid, CoA, FAD, and NAD^+. This process catalyses the transfer of a 2-carbon unit from a 2-keto sugar to C-1 of various aldoses.

385. The answer is E. *(White, ed 5. p 732.)* Folic acid as tetrahydrofolate tranfers methyl groups. Its structural analog competitor, methotrexate, prevents the reduction of dihydrofolate to tetrahydrofolate, by interference with the enzyme dihydrofolate reductase.

386. The answer is B. *(Harper, ed 15. pp 106-107.)* Pantothenic acid, also called co-acetylase, is a component of CoA. Acetyl CoA is the activated form of acetate employed in acetylation reactions, including the Krebs cycle, and lipid and cholesterol metabolism.

387. The answer is C. *(Harper, ed 15. p 55.)* The compound whose diagram is shown in the question is S-adenosylmethionine. It functions in many biochemical reactions as a methyl donor.

388. The answer is A. *(Harper, ed 15. pp 254-255.)* The cofactors involved in the pyruvate dehydrogenase reaction include thiamine pyrophosphate, lipoic acid, FAD, NAD^+, and CoA.

389. The answer is E. *(Harper, ed 15. p 112.)* Raw egg white contains a protein, avidin, which complexes specifically with biotin and has been shown to induce biotin deficiency in laboratory animals. It was by means of avidin that the factor biotin was first shown to be a growth factor.

390. The answer is C. *(Mahler, ed 2. pp 422-424.)* Biotin is involved in carboxylation reactions such as those catalyzed by acetyl CoA, pyruvate, and propionyl CoA carboxylases. It is best known in connection with CO_2 fixation.

391. The answer is A. *(Harper, ed 15. pp 112-113.)* Biotin is a cofactor of many carboxylase enzymes. Succinic thiokinase, however, is not a carboxylase, does not require biotin and, therefore, is not affected by avidin.

392. The answer is D. *(Harper, ed 15. pp 326-328.)* Pyridoxal phosphate is an obligatory cofactor for the reactions of transaminases, and of many other enzymes with amino acid substrates. In these reactions it forms a Schiff base intermediate which then can be rearranged, transaminated, etc.

393. The answer is C. *(Orten, ed 8. p 819.)* Folic acid is synthesized from a pteridine nucleus, *p*-aminobenzoic acid, and glutamate.

394. The answer is B. *(Harper, ed 15. pp 106-107.)* Pantothenate is the precursor of coenzyme A, and together they partake in numerous reactions throughout the metabolic scheme. There is no documented deficiency state for pantothenate.

395. The answer is C. *(Harper, ed 15. pp 94-96.)* Vitamin E is a tocopherol. As a fat-soluble vitamin, its deficiency occurs in patients with fat malabsorption. Deficiency in some species of animals is associated with gonadal dysfunction, anemia, and skin changes.

396. The answer is A. *(White, ed 5. pp 1162-1163.)* Pellagra is thought to result chiefly from a deficiency of nicotinic acid (niacin). Photodermatitis, stomatitis, glossitis, and diarrhea, with central neurologic changes are found in pellagra, and alleviated by niacin.

397. The answer is E. *(White, ed 5. pp 1157-1160.)* Deficiency of riboflavin may cause fissuring at the corners of the mouth and lips (cheilosis). (Cheilosis is more commonly seen with iron deficiency and with dentures.) Riboflavin deficiency is also associated with corneal vascularization, seborrheic dermatitis, and glossitis.

398. The answer is B (1, 3). *(White, ed 5. pp 1101-1103.)* Glucagon accelerates glycogenolysis in the liver by activating phosphorylase kinase and, consequently, liver phosphorylase. Skeletal muscle is not a target for glucagon action. Gluconeogenesis is accelerated, not depressed by glucagon.

399. The answer is B (1, 3). *(Mahler, ed 2. pp 47-50, 66, 635.)* Prostaglandins are derived from homo-γ-linolenic acid (a fatty acid). Vasopressin is a nonapeptide secreted by the pituitary. Thyroxine and epinephrine are derived from tyrosine.

400. The answer is B (1, 3). *(Mahler, ed 2. pp 168-170.)* Insulin, which consists of two different polypeptide chains held together by disulfide bonds, is synthesized by cleaving a "connecting peptide" out of a larger molecule called proinsulin. When proinsulin is denatured, it can be renatured to give native molecules because the primary structure of the polypeptide chain determines its higher order structure. In contrast, because of the absence of the connecting peptide, the primary structure of insulin does not determine its higher order structure. After denaturation, insulin cannot be renatured successfully in good yield.

401. The answer is A (1, 2, 3). *(Williams, ed 5. pp 40-46.)* The hormones HCG, LH, TSH and FSH are all glycoprotein hormones which contain two polypeptide chains. The alpha-chains are common to all four hormones while the beta-chains differ.

402. The answer is E (all). *(Mahler, ed 2. pp 513, 545, 803, 894.)* The vitamin shown is nicotinic acid. NAD^+ is a cofactor required by lactate dehydrogenase, UDP glucose epimerase, and polynucleotide ligase of *E. coli*. NADPH is a cofactor for phenylalanine hydroxylase.

403. The answer is A (1, 2, 3). *(Mahler, ed 2. pp 401-406.)* The vitamin shown is thiamine. Thiamine pyrophosphate is required for the reactions catalyzed by pyruvate dehydrogenase, transketolase, and α-ketoglutarate dehydrogenase. In all of these reactions, thiamine is involved with oxidative decarboxylation. Glutamate dehydrogenase neither involves oxidative decarboxylation nor thiamine.

404. The answer is A (1, 2, 3). *(Mahler, ed 2. pp 393-401.)* The vitamin shown is pyridoxine (vitamin B_6). Pyridoxal phosphate is required by glutamic-oxaloacetic transaminase, glycogen phosphorylase, and dihydroxyphenylalanine decarboxylase.

405. The answer is C (2, 4). *(Mahler, ed 2. pp 414-418, 519-520.)* The vitamin shown is riboflavin. The cofactor FAD is involved in the reactions catalyzed by pyruvate dehydrogenase and succinate dehydrogenase.

406. The answer is' A (1, 2, 3). *(Harper, ed 15. pp 100-101.)* Thiamine pyrophosphate is a cofactor for such enzymes as transketolase as well as for decarboxylases of α- keto acids. Thiamine pyrophosphate is a cofactor for both oxidative and non-oxidative decarboxylation of such molecules as pyruvate.

407. The answer is A (1, 2, 3). *(White, ed 5. pp 1152-1200.)* Vitamin A (retinol), vitamin D (calciferol), and vitamin E (tocopherol) are all fatsoluble vitamins. Ascorbic acid, vitamin C, is a water-soluble vitamin, as are the B vitamins as a group, and folate.

408. The answer is E (all). *(Mahler, ed 2. pp 393-401.)* Pyridoxal phosphate is a cofactor in many types of reactions: decarboxylation (glutamate decarboxylase); deamination (serine dehydratase); transamination (glutamic-oxaloacetic transaminase); and transsulfuration (cystathionine synthetase and cystathionase).

409. The answer is B (1, 3). *(White, ed 5. pp 1042, 1079, 1112.)* Parathormone is secreted by the parathyroid glands and is involved in calcium and phosphate homeostasis. Aldosterone is a mineralocorticoid produced by the adrenal gland. Only ACTH and FSH are produced by the adenohypophysis.

410. The answer is D (4). *(White, ed 5. pp 173-176, 1152-1200.)* Vitamin B_{12}, a complex molecule containing cobalt and a porphyrin-like ring system, is a water-soluble vitamin. The B vitamins are generally water-soluble, whereas vitamins A, D, E, and K are all fat-soluble.

411. The answer is **A (1, 2, 3).** *(White, ed 5. pp 334, 336, 339.)* FAD, CoA, and NAD contain, respectively, riboflavin, pantothenic acid, and nicotinic acid, all of which are vitamins. ATP is composed of adenosine, ribose, and phosphate only.

412. The answer is **B (1, 3).** *(Mahler, ed 2. pp 406-409.)* The vitamin shown is folic acid. Tetrahydrofolate is involved as a donor of one-carbon units in the biosynthesis of ATP and TTP. In the latter case, tetrahydrofolate also serves as a source of reducing equivalents.

413. The answer is **B (1, 3).** *(Mahler, ed 2. pp 418-422, 424-427, 656-657.)* Hemoglobin and cytochrome c contain heme cofactors that are porphyrin derivatives. Methylmalonyl CoA mutase requires vitamin B_{12} which contains a corrin ring resembling, but different from, a porphyrin ring. Ferredoxin contains nonheme iron.

414. The answer is **A (1, 2, 3).** *(Williams, ed 5. p 504.)* The six-amino acid ring is a marked structural characteristic of insulin, oxytocin, and vasopressin and may be a physical property of the molecule that is involved in binding to specific receptor sites on target cells.

415. The answer is **D (4).** *(Sawin, 1969. pp 4-6.)* Since actinomycin D blockage of RNA synthesis, and mitomycin C blockage of DNA synthesis, did not inhibit the action of the hormone, then DNA synthesis and RNA synthesis were not required. Puromycin and cycloheximide blocked the hormone through their action as inhibitors of protein synthesis.

416-418. The answers are: **416-A, 417-C, 418-E.** *(White, ed 5. pp 1023, 1042, 1072.)* ACTH is a peptide hormone produced by the adenohypophysis. It stimulates the synthesis of steroids by the adrenal cortex. The neurohypophysis produces vasopressin and oxytocin. Both epinephrine and norepinephrine are produced in the adrenal medulla. Steroid hormones are produced by the adrenal cortex. The parafollicular cells of the thyroid gland produce calcitonin which regulates Ca^{++} metabolism. These cells are the origin of medullary carcinoma of the thyroid, in which calcitonin assay provides a specific means of following the tumor's course.

419-421. The answers are: **419-E, 420-A, 421-B.** *(White, ed 5. pp 915-919.)* Secretin is a circulatory hormone liberated in response to peptides or acid in the duodenum. It causes an increased flow of pancreatic juice. Gastrin governs acid production by the stomach, and cholecystokinin causes gallbladder contraction. Cholecystokinin is released by the duodenum into the circulation to stimulate this contraction, causing subsequent emptying of bile into the intestine. Its C-terminal octapeptide is more than five times as potent as the parent hormone, and its C-terminal pentapeptide is identical with gastrin. Gastrin is produced in the specific cells of the antral mucosa of the stomach. It stimulates parietal cells to produce HCl, approximately 0.16 M, with KCl 0.007 M. It also stimulates glucagon and insulin secretion. Gastrin production is inhibited by secretin.

422-424. The answers are: **422-B, 423-C, 424-E.** *(White, ed 5. pp 910-915.)* Pancreatic juice contains approximately 80 mEq/l of HCO_3^-, which neutralizes the acidity of gastric secretions. It contains appreciable quantities of potassium, as well. The parietal cells of the stomach secrete 0.16 M HCl and other trace electrolytes. These are stimulated by neural (vagal) and by hormonal (gastrin) mechanisms. Pharmacologically, stimulation by histamine provides a routine provocative test used in medicine. Tears contain the enzyme lysozyme which can protect the eye from infection by hydrolyzing the cell walls of many bacteria.

425-427. The answers are: **425-A, 426-C, 427-D.** *(Wintrobe, ed 7. pp 1948-1950.)* 7-Dehydrocholesterol is converted into cholecalciferol (vitamin D_3) photochemically in the skin. Cholecalciferol is hydroxylated at the 25-position in liver and subsequently at the 1-position in kidney.

Membranes and Cell Structure

428. The answer is C. *(Mahler, ed 2. pp 433-434.)* The structure shown is a mitochondrion, characterized by a smooth outer membrane and a complex inner membrane with many folds called cristae.

429. The answer is A. *(Mahler, ed 2. pp 433-434.)* The structure shown is a mitochondrion which is the energy producing organelle of the cell. Lysosomes are the degradative machinery of the cell, and Golgi bodies are engaged in the packaging of secretory proteins. Ribosomes accomplish protein synthesis and DNA accomplishes eukaryotic cell genetic transfer.

430. The answer is C. *(Wolfe, 1972. p 346.)* Microtubules, as shown, are hollow, tubular structures 100 Å in diameter composed of two protein subunits. They have a structural function in the cell, and are associated with the movement of chromosomes and the flow of neurotransmitters in axons.

431. The answer is B. *(Mahler, ed 2. pp 457-458.)* A typical plasma membrane contains 35-40 percent lipid, 55 percent protein, up to 5 percent carbohydrate, and less than 0.1 percent RNA by weight.

432. The answer is D. *(White, ed 5. pp 45, 415-416.)* Phlorhizin is an inhibitor of sodium-dependent glucose transport, while phloretin inhibits sodium-independent facilitated diffusion. They are both glycosides, and related to the digitalis glycoside ouabain, which inhibits Na^+-K^+-ATPase which ejects Na^+ from cells.

433. The answer is D. *(Wolfe, 1972. p 104.)* The mitochondrion, as shown, is an intracellular body typically 0.5 μ by 2.0 μ in size. It contains a substructure of cristae, granular matrix, and DNA, and is involved in the production of energy for the cell.

434. The answer is C. *(Wolfe, 1972. p 20.)* The illustrated junction is a desmosome. Desmosomes are dense plaques consisting of specialized fibrils which are found between epithelial cells, serving to attach the membranes of adjacent cells.

435. The answer is C (2, 4). *(White, ed 5. pp 542-578.)* The chain elongation of fatty acids synthesized de novo, including palmitic and stearic acid, occurs in the mitochondria. The elongation of polyunsaturated fatty acids occurs in the soluble fraction of the cell and requires biotin.

436. The answer is B (1, 3). *(Mahler, ed 2. pp 457-459.)* The structure shown is a plasma membrane composed of protein and lipid bilayers. It contains a Na^+/K^+ pump which is dependent on an ATPase enzyme. Cellular energy is produced primarily in mitochondria.

437. The answer is A (1, 2, 3). *(Mahler, ed 2. pp 457-458.)* Crucial features of the Davson-Danielli model of membrane structure include an arrangement of membrane lipids in a bilayer with hydrophobic side chains pointing inward. The hydrophilic heads, which are directed outward, interact with a monolayer of protein which in turn interacts with the aqueous environment.

438. The answer is B (1, 3). *(Lehninger, 1970. p 31.)* Protein synthesis occurs in the cytoplasm, on groups of ribosomes called polysomes, and on ribosomes associated with membranes, termed the rough endoplasmic reticulum.

439. The answer is E (all). *(Mahler, ed 2. p 452.)* Lysosomes are cellular organelles which degrade ingested particles. Cathepsins are proteases which, like the rest of the enzymes listed, breakdown biochemicals in lysosomes.

440. The answer is D (4). *(Bloom, ed 9. pp 112-116.)* Lysed red blood cell ghosts contain little hemoglobin. They are well-studied preparations of intact membrane bilayers with the usual content of cholesterol, and no mitochondria.

441-444. The answers are: 441-A, 442-B, 443-C, 444-D. *(Bloom, ed 9. pp 37-60. Wolfe, 1972. p 16.)* The subcellular organelles listed are: 1) a bilayer membrane with or without attached ribosomes in the cytosol; 2) a continuity between the inner and outer nuclear membranes; 3) a membrane-bound body containing cristae, a finely granular matrix, and elementary bodies; 4) the limiting bilayer membrane of the cell.

445-447. The answers are: 445-A, 446-C, 447-B. *(Mahler, ed 2. pp 432-438.)* The nucleus is the site of both DNA replication and RNA transcription. DNA remains in the nucleus, while mRNA is transported to the cytoplasm to direct protein synthesis. The mitochondria are the sites of the electron transport chain which ultimately oxidizes molecular oxygen and, in its progression, is coupled to the phosphorylation of ADP to ATP. The endoplasmic reticulum consists of lipoprotein membranes, and is divided into smooth and rough types. The smooth endoplasmic reticulum lacks ribosomes; the rough endoplasmic reticulum is lined with ribosomes and accomplishes protein synthesis.

Metabolism

448. The answer is C. *(White, ed 5. p 454.)* The Pasteur effect is, solely, the observation that the rate of glucose catabolism (glycolysis) is decreased under aerobic conditions, relative to anaerobic conditions. The observation was initially made with microorganisms, but holds true with mammalian tissues and with tumors. It is explained by O_2 effects on phosphofructokinase.

449. The answer is D. *(White, ed 5. pp 836-838, 847.)* The affinity of hemoglobin for oxygen is decreased by an increase in the CO_2 tension. This is attributable solely to change in pH, and is known as the Bohr effect.

450. The answer is E. *(White, ed 5. pp 879-904.)* In metabolic acidosis, blood bicarbonate is found to be low. In a situation of metabolic acidosis where bicarbonate has been consumed, compensation can be made by the respiratory mechanism of hyperventilation, so that pH is normalized by reducing plasma carbon dioxide.

451. The answer is D. *(White, ed 5. pp 100-102, 889.)* According to the Henderson-Hasselbalch equation, pH = pKa + $\log \frac{\text{base}}{\text{acid}}$. In the case of 0.2 M lactate and 0.02 M lactic acid, pH = 3.9 + log 10 = 4.9.

452. The answer is B. *(White, ed 5. pp 951-953.)* The ATPase which hydrolyzes ATP, thereby releasing the energy that allows for muscle contraction, is located on the myosin molecule. The other listed components of muscle lack ATPase activity.

453. The answer is D. *(Davis, ed 2. pp 600, 606.)* The A antigen is a surface glycoprotein, and part of the ABO system. The Kell system is a separate red blood cell antigen system, and does not react with the A antigen.

454. The answer is A. *(Sutherland, E W. Diabetes. 18:797, 1969. White, ed 5. pp 288-290.)* Cyclic AMP phosphodiesterase is the intracellular enzyme that degrades cyclic AMP to 5′-AMP. Adenylate cyclase synthesizes cyclic AMP from ATP. The remaining compounds inhibit phosphodiesterase and would increase the levels of cyclic AMP.

455. The answer is E. *(Mahler, ed 2. p 38.)* All of the compounds listed contain high-energy phosphate bonds and are found in muscle tissue, but creatine phosphate is the molecular "energy storehouse" of muscle cells.

456. The answer is C. *(Mahler, ed 2. p 13.)* The maximum buffering capacity is at a pH equal to the pKa of the acid. Physiologic pH is about 7.4, so that NaH_2PO_4 is the most effective buffer among those listed.

457. The answer is B. *(Lehninger, 1970. p 43.)* The dipole nature of the water molecule allows it to dissolve salts by electrostatic interaction, and non-ionic compounds by hydrogen bonding with polar functional groups.

458. The answer is C. *(White, ed 5. pp 126-128, 880.)* For an albumin content of 4.4 g/100 ml and a globulin content of 2.63 g/100 ml, (normal), the calculated osmotic pressure is 27.9 mm Hg. This is empirically the pressure range in which edema, or loss of intravascular fluid into the tissues (pulmonary edema), is seen.

459. The answer is D. *(White, ed 5. p 805.)* Albumin is responsible for 75 to 80 percent of the osmotic effect of serum proteins. This is because it is the smallest and most abundant of the plasma proteins, though by weight it usually only comprises slightly more than half of the plasma proteins.

460. The answer is B. *(Mahler, ed 2. pp 690-701.)* In tightly coupled mitochondria, respiration can only proceed in the presence of phosphate plus phosphate acceptor (ADP), as well as respiratory substrate. 2,4-Dinitrophenol uncouples respiration from phosphorylation thus allowing active respiration in the absence of ADP. Oligomycin inhibits the ATPase responsible for catalyzing the phosphorylation of ADP. In mitochondria previously uncoupled by 2,4-dinitrophenol, oligomycin does not affect respiration.

461. The answer is B. *(Lehninger, 1970. p 209.)* Lipid micelles are stable in water when the polar heads of the lipid face outward in contact with water and the hydrophobic, nonpolar tails turn inward to exclude water.

462. The answer is A. *(White, ed 5. p 817.)* Elevated levels of acid phosphatase in serum characteristically are a sign of metastatic carcinoma of the prostate. Alkaline phosphatase elevations typically mark the other conditions listed.

463. The answer is E. *(Bernhard, H P. Dev Biol. 35:83-96, 1973.)* Gluconeogenesis from certain substrates and the hydroxylation of phenylalanine can be performed by kidney as well as liver, and glycogen synthesis and storage are carried out by muscle as well as liver. A wide variety of cell types are sensitive to epinephrine. Serum albumin synthesis is a known organotypic function of liver.

464. The answer is B. *(Brown, M S. Proc Natl Acad Sci. 70:2162-2166, 1973.)* The uptake of exogenous cholesterol by cells results in a marked suppression of endogenous cholesterol synthesis. Human, low-density lipoprotein not only contains the greatest ratio of bound cholesterol to protein, but also has the greatest potency in suppressing endogenous cholesterogenesis.

465. The answer is D. *(Gilbert, F. Proc Natl Acad Sci. 72:263-267, 1975. White, ed 5. p 589.)* In Sandhoff's disease, both the A and B forms of hexosaminidase are absent. Hexosaminidase A-deficiency is seen in Tay-Sachs disease. The other deficiency states described are not now assigned to specific syndromes.

466. The answer is A. *(Fujimoto, WY. Proc Natl Acad Sci. 68:1516-1519, 1971. White, ed 5. p 768.)* The Lesch-Nyhan syndrome is a sex-linked, recessive disorder. Infantile autism and Cushing's syndrome do not have an inherited transmission. Maple syrup urine disease is autosomal recessive, and familial hypercholesterolemia is codominant.

467. The answer is C. *(White, ed 5. pp 127-129.)* There are two conditions to be satisfied in the Donnan equilibrium: 1) electroneutrality of solutions on both sides of the membrane; and 2) chemical gradients for oppositely charged permeations must be equal and opposite. Thus, on the left side $(Na)_l = (Prot) + (Cl)_l = 100 + (Cl)_l$. On the right side, $(Na)_r = (Cl)_r$. Furthermore, $(Na)_l/(Na)_r = (Cl)_r/(Cl)_l$. Finally, according to conservation of mass, $(Na)_l + (Na)_r = 200$ and $(Cl)_l + (Cl)_r = 100$. All of these conditions are satisfied for the following:

$(Na)_l = 133\frac{1}{3}$ $(Na)_r = 66\frac{2}{3}$
$(Cl)_l = 33\frac{1}{3}$ $(Cl)_r = 66\frac{2}{3}$

468. The answer is A. *(Bondy, ed 7. p 805.)* Porphobilinogen is colorless and so has no effect on the color of urine when first excreted. On standing at an acid pH, the urine containing colorless porphobilinogen may polymerize and oxidize to form porphyrins and other pigments, and so darken.

469. The answer is C. *(Bondy, ed 7. pp 123, 316-317.)* Muscle phosphorylase deficiency leads to a glycogen storage disease, and the inability to do strenuous muscular work in ischemic situations. This is because muscular activity in such situations is dependent upon ambient fatty acids and glucose.

470. The answer is D. *(White, ed 5. pp 1001-1005.)* Alkaline phosphatase, which is present in unusually high activity in osteoblasts, is thought to play a role in producing a local increase in the phosphate concentration. Phosphorylase kinase is involved with glycogen synthesis, and glucose 6-phosphate dehydrogenase is involved with the phosphogluconate pathway. Acid phosphatase is seen chiefly in the prostate, whereas ATPase is a widespread enzyme involved in membrane transport and a variety of other reactions.

471. The answer is A. *(Bondy, ed 7. p 596.)* In phenylketonuria, the enzyme phenylalanine hydroxylase which synthesizes tyrosine from phenylalanine is absent. This leads to a classic syndrome of eczema, light pigmentation, seizures, and mental retardation which can be prevented by early detection of the deficiency.

472. The answer is B. *(Bondy, ed 7. pp 314-315.)* Von Gierke's disease (type I glycogen storage disease) results from a deficiency of glucose 6-phosphatase. Phosphoglucomutase deficiency is seen in type VII glycogen storage disease. The other cited enzymes are not specifically involved in glycogenoses.

473. The answer is E. *(White, ed 5. p 794.)* Phenylketonuria results from phenylalanine hydroxylase deficiency. Tay-Sachs disease is caused by accumulation of excess gangliosides due to enzymatic deficiencies. Glucose 6-phosphatase deficiency leads to von Gierke's disease, and enzyme deficiencies result in accumulation of glucocerebrosides in Gaucher's disease. Caisson disease, or decompression sickness, is unrelated to an enzyme defect or deficit.

474. The answer is C. *(Bondy, ed 7. pp 314-316.)* Type IV glycogen storage disease (Andersen's disease) results from a lack of the brancher enzyme amylo-(1,4→1,6)-transglycosylase. The defect results in an abnormal glycogen structure with very long unbranched chains which can be detected in the liver, on biopsy. Hepatomegaly and cirrhosis occur in this condition, which is usually fatal before the age of three years.

475. The answer is D. *(White, ed 5. pp 910-912.)* Parietal cells in the gastric mucosa secrete hydrochloric acid, lowering the gastric contents to a pH of 1-2. The acidity of gastric juice is neutralized in the intestine by alkaline pancreatic and biliary juice to pH 7-8. The pH of plasma is normally 7.4, while sweat is composed largely of half-normal saline, at a pH of 4.5-7.5.

476. The answer is C (2, 4). *(Mahler, ed 2. pp 489-491.)* Although the same intermediates may appear in both anabolic and catabolic pathways, one path is not simply the reverse of the other, because irreversible enzymatic steps often occur in the beginning of the reaction sequence. The same enzymes often appear in innumerable metabolic pathways.

477. The answer is A (1, 2, 3). *(Wintrobe, ed 7. pp 172-173.)* The affinity of hemoglobin for oxygen is decreased by such substances as 2,3-diphosphoglycerate, carbon dioxide, and hydrogen ions.

478. The answer is D (4). *(Harper, ed 15. p 315.)* During starvation, the level of gluconeogenic enzymes increases in the liver. Enzymes of the pentose cycle, glycolysis, and lipogenesis decrease in activity during starvation, as would be expected, since substrates for these pathways are then lacking.

479. The answer is C (2, 4). *(White, ed 5. pp 1136, 1169, 1179-1180.)* Biotin and dehydroshikimic acid are not synthesized in mammalian tissue. While inositol is required in the diet, isotopic evidence suggests that at least small amounts are synthesized in mammalian tissue. Pathways also exist in mammalian tissues for the synthesis of choline via phosphatidyl choline.

480. The answer is E (all). *(Wintrobe, ed 7. p 1518.)* Alkaline phosphatase is a ubiquitous enzyme in humans. Conditions in which any of the tissues listed are involved can lead to increased serum alkaline phosphatase.

481. The answer is D (4). *(Wintrobe, ed 7. pp 397-399.)* While erythrocytes can metabolize glucose anaerobically to lactate via the glycolytic pathway, this process does not involve any net oxidation of glucose. The NADH derived from the triose phosphate dehydrogenase step is used to reduce pyruvate to lactate. Liver and heart are widely adaptable in their metabolic requirements, and can oxidize glucose well. The brain's cells are almost exclusively dependent upon glucose for oxidative metabolism and energy production.

482. The answer is C (2, 4). *(McGilvery, 1970. pp 272-280.)* Liver and kidney are the two important sites of gluconeogenesis in mammals: neither red cells nor skeletal muscle have any role in gluconeogenesis.

483. The answer is B. *(White, ed 5. p 692.)* The metabolic block in tyrosinosis appears to be at the level of p-hydroxyphenylpyruvic acid oxidase with consequent excessive excretion of p-hydroxyphenylpyruvate.

484. The answer is D (4). *(Mahler, ed 2. p 857.)* Phenylketonuria is thought to result from a genetic deficiency in the enzyme phenylalanine hydroxylase which converts phenylalanine to tyrosine. The resulting dementia, seizures, and morbidity are largely preventable, on the basis of perinatal screening and treatment.

485. The answer is B (1, 3). *(Bondy, ed 7. pp 578-579.)* Maple syrup urine disease results from a deficiency of the α-ketoacid dehydrogenases responsible for oxidation of the α-keto analogs of the branched chain amino acids (leucine, isoleucine, and valine). Consequently, these α-ketoacids accumulate in blood, urine, and spinal fluid producing the characteristic pathology and odor.

486. The answer is C (2, 4). *(Bondy, ed 7. pp 618, 620.)* On standing, porphobilinogen and homogentisic acid will turn urine black as seen in acute intermittent porphyria and ochronosis, respectively. Metanephrine and 5-hydroxyindoleacetic acid are colorless.

487. The answer is A (1, 2, 3). *(Bondy, ed 7. pp 315-316.)* The first three enzymes listed help to break down glycogen to glucose. The fourth enzyme is necessary for the synthesis of glycogen.

488. The answer is E (all). *(Bondy, ed 7. pp 251, 262-263.)* All of the diseases mentioned are known to be associated with abnormal glucose tolerance test curves in some patients. Mechanisms for glucose intolerance in these cases can include post-surgical rapid transit in the gastrointestinal tract, steroid therapy, and ectopic ACTH production.

489. The answer is B (1, 3). *(Lynch, ed 2. p 387.)* Serum insulin levels cannot be measured accurately once chronic insulin therapy is begun, since antibodies that interfere with the measurement made by immunoassay will be formed. Insulin requirements may decrease to nothing for a short period termed the "honeymoon phase." Impotence is a variable, but usually late, complication associated with neuropathy. Diet governs insulin requirement and, therefore, is the mainstay of "regulation" in the insulin-requiring, as well as in the non-insulin requiring, diabetic patient.

490. The answer is E (all). *(Beeson, ed 14. pp 711, 1348, 1391.)* Alcohol abuse may lead to all of the complications listed, either directly or in association with malnutrition.

491. The answer is C (2, 4). *(McGilvery, 1970. pp 369-373, 526-527.)* Ketone bodies can be oxidized to carbon dioxide in virtually all tissues which contain mitochondria (brain, kidney, heart, skeletal muscle, etc.), but not in the liver. Cells without mitochondria, such as erythrocytes, cannot oxidize ketone bodies.

492. The answer is D (4). *(White, ed 5. pp 576-578, 975.)* Ketone bodies are synthesized chiefly in liver, and consist of β-hydroxybutyrate, aceto-acetate, and acetone.

493-496. The answers are: 493-A, 494-C, 495-E, 496-D. *(Orten, ed 8. p 102.)* Tryptophan pyrrolase is deficient in Hartnup's disease. Uridine diphosphate glucuronate transferase is lacking in the Crigler-Najjar syndrome. Ceruloplasmin is deficient in Wilson's disease. Pyruvate kinase deficiency leads to a hemolytic anemia, and albinism may be caused by tyrosinase deficiency.

497-500. The answers are: 497-A, 498-A, 499-A, 500-D. *(Bondy, ed 7. pp 364-367.)* In patients with juvenile diabetes, endogenous insulin production is absent to a degree not seen in adult onset diabetes; thus, there is an absolute need for insulin. These patients have a greater tendency to develop ketoacidosis, and their diabetes is considerably more unstable and "brittle." Insulin administration to diabetic patients over the course of several months causes the development of antibodies to insulin. However, insulin resistance is rare in both juvenile and adult onset forms.

Bibliography

Anderson, W. A., ed. *Pathology.* 6th rev. ed. 2 vols. St. Louis: C. V. Mosby Co., 1971.

Beeson, P. B., and McDermott, W., eds. *Cecil-Loeb Textbook of Medicine.* 14th ed. 2 vols. Philadelphia: Columbia Broadcasting System, W. B. Saunders Co., 1975.

Bernhard, H. P., et al. *Dev Biol* 35 (1973):83-96.

Bloom, W., and Fawcett, D. W. *Textbook of Histology.* 9th ed. Philadelphia: Columbia Broadcasting System, W. B. Saunders Co., 1968.

Bondy, P. K., and Rosenberg, L. E. *Genetics & Metabolism.* vol. 1. *Endocrinology.* vol. 2. Duncan's Diseases of Metabolism: The Genetic & Biochemical Basis of Disease. 7th ed. Philadelphia: Columbia Broadcasting System, W. B. Saunders Co., 1974.

Brown, M. S., et al. *Proc Natl Acad Sci* 70 (1973):2162-2166.

Davis, B. D.; Dulbecco, R.; Eisen, H. N.; Ginsberg, H. S.; and Wood, W. B. *Microbiology.* 2nd ed. New York: Harper & Row Publications, Inc., Harper Medical Department, 1973.

Fujimoto, W. Y., et al. *Proc Natl Acad Sci* 68 (1971):1516-1519.

Gilbert, F., et al. *Proc Natl Acad Sci* 72 (1975):263-267.

Harper, H. A. *Review of Physiological Chemistry.* 15th ed. Los Altos: Lange Medical Publications, 1975.

Hayes, W. *Genetics of Bacteria and Their Viruses.* 2nd ed. New York: John Wiley & Sons, Inc., Halsted Press, 1969.

Jawetz, E.; Melnick, J. L.; and Adelberg, E. A. *Review of Medical Microbiology.* 11th ed. Los Altos: Lange Medical Publications, 1974.

Jelinek, W., et al. *J Mol Biol* 75 (1973):515

Lehninger, A. *Biochemistry.* New York: Worth Publishers, Inc., 1970.

Lewin, B. *Bacterial Genomes.* Gene Expression. vol. 1. New York: John Wiley & Sons, Inc., Wiley-Interscience, 1974.

Lindros, K. O. et al. *J Biol Chem* 249 (1974):7956-7963.

Lynch, M. J.; Raphael, S. S.; Mellor, L. D.; Spare, P. D.; and Inwood, M. J. H. *Medical Laboratory Technology and Clinical Pathology.* 2nd ed. Columbia Broadcasting System, W. B. Saunders Co., 1969.

McGilvery, R. W. *Biochemistry: A Functional Approach.* Philadelphia: Columbia Broadcasting System, W. B. Saunders, 1970.

Mahler, H. R., and Cordes, E. H. *Basic Biological Chemistry.* 2nd ed. New York: Harper & Row Publications, Inc., Harper College Books, 1971.

Miller, J. H. *Experiments in Molecular Genetics.* New York: Cold Spring Harbor Laboratory, 1972.

Orten, J. M., and Neuhaus, O. W. *Biochemistry.* 8th ed. St. Louis: C. V. Mosby., 1970.

Sawin, C. T. *Hormones: Endocrine Physiology.* Boston: Little, Brown & Co., 1969.

Sutherland, E. W., et al. *Diabetes* 18 (1969):797.

Watson, J. D. *Molecular Biology of the Gene.* 2nd ed. New York: W. A. Benjamin Co., Inc., 1970.

White, A.; Handler, P.; and Smith, E. L. *Principles of Biochemistry.* 5th ed. New York: McGraw-Hill Book Co., Blakiston Publications, 1973.

Williams, R. H., ed. *Textbook of Endocrinology.* 5th ed. Philadelphia: Columbia Broadcasting System, W. B. Saunders Co., 1974.

Wintrobe, M. M.; Thorn, G. W.; Raymond, D. A.; Braunwald, E.; Isselbacher, K. J.; and Petersdorf, R. G., eds. *Harrison's Principles of Internal Medicine.* 7th rev. ed. New York: Mc-Graw-Hill Book Co., Blakiston Publications, 1974.

Wolfe, S. *Biology of the Cell.* Belmont: Wadsworth Publications, 1972.

NOTES

NOTES

NOTES